Convolutional Calculus

Mathematics and Its Applications (*East European Series*)

Volume 43

Convolutional Calculus

by

Ivan H. Dimovski
Institute of Mathematics,
Bulgarian Academy of Sciences,
Sofia, Bulgaria

KLUWER ACADEMIC PUBLISHERS
DORDRECHT / BOSTON / LONDON

Library of Congress Cataloging in Publication Data

Dimovski, I.
Convolutional calculus / by Ivan H. Dimovksi. -- [2nd ed.]
p. cm. -- (Mathematics and its applications. East European series ; 43)
Includes bibliographical references.
ISBN-13:978-94-010-6723-2 e-ISBN-13:978-94-009-0527-6
DOI: 10.1007/978-94-009-0527-6
1. Mathematical analysis. 2. Convolutions (Mathematics)
3. Multipliers (Mathematical analysis) 4. Linear operators.
I. Title. II. Series: Mathematics and its applications (Kluwer Academic Publishers). East European series ; 43.
QA300.D48 1990
515--dc20 89-26851

ISBN-13:978-94-010-6723-2

Published by Kluwer Academic Publishers,
P.O. Box 17, 3300 AA Dordrecht, The Netherlands.

Kluwer Academic Publishers incorporates
the publishing programmes of
D. Reidel, Martinus Nijhoff, Dr W. Junk and MTP Press.

Sold and distributed in the U.S.A. and Canada
by Kluwer Academic Publishers,
101 Philip Drive, Norwell, MA 02061, U.S.A.

In all other countries, sold and distributed
by Kluwer Academic Publishers Group,
P.O. Box 322, 3300 AH Dordrecht, The Netherlands.

Printed on acid-free paper

SERIES EDITOR'S PREFACE

'Et moi, ..., si j'avait su comment en revenir, je n'y serais point allé.'

Jules Verne

The series is divergent; therefore we may be able to do something with it.

O. Heaviside

One service mathematics has rendered the human race. It has put common sense back where it belongs, on the topmost shelf next to the dusty canister labelled 'discarded non-sense'.

Eric T. Bell

Mathematics is a tool for thought. A highly necessary tool in a world where both feedback and non-linearities abound. Similarly, all kinds of parts of mathematics serve as tools for other parts and for other sciences.

Applying a simple rewriting rule to the quote on the right above one finds such statements as: 'One service topology has rendered mathematical physics ...'; 'One service logic has rendered computer science ...'; 'One service category theory has rendered mathematics ...'. All arguably true. And all statements obtainable this way form part of the raison d'être of this series.

This series, *Mathematics and Its Applications*, started in 1977. Now that over one hundred volumes have appeared it seems opportune to reexamine its scope. At the time I wrote

> "Growing specialization and diversification have brought a host of monographs and textbooks on increasingly specialized topics. However, the 'tree' of knowledge of mathematics and related fields does not grow only by putting forth new branches. It also happens, quite often in fact, that branches which were thought to be completely disparate are suddenly seen to be related. Further, the kind and level of sophistication of mathematics applied in various sciences has changed drastically in recent years: measure theory is used (non-trivially) in regional and theoretical economics; algebraic geometry interacts with physics; the Minkowsky lemma, coding theory and the structure of water meet one another in packing and covering theory; quantum fields, crystal defects and mathematical programming profit from homotopy theory; Lie algebras are relevant to filtering; and prediction and electrical engineering can use Stein spaces. And in addition to this there are such new emerging subdisciplines as 'experimental mathematics', 'CFD', 'completely integrable systems', 'chaos, synergetics and large-scale order', which are almost impossible to fit into the existing classification schemes. They draw upon widely different sections of mathematics."

By and large, all this still applies today. It is still true that at first sight mathematics seems rather fragmented and that to find, see, and exploit the deeper underlying interrelations more effort is needed and so are books that can help mathematicians and scientists do so. Accordingly MIA will continue to try to make such books available.

If anything, the description I gave in 1977 is now an understatement. To the examples of interaction areas one should add string theory where Riemann surfaces, algebraic geometry, modular functions, knots, quantum field theory, Kac-Moody algebras, monstrous moonshine (and more) all come together. And to the examples of things which can be usefully applied let me add the topic 'finite geometry'; a combination of words which sounds like it might not even exist, let alone be applicable. And yet it is being applied: to statistics via designs, to radar/sonar detection arrays (via finite projective planes), and to bus connections of VLSI chips (via difference sets). There seems to be no part of (so-called pure) mathematics that is not in immediate danger of being applied. And, accordingly, the applied mathematician needs to be aware of much more. Besides analysis and numerics, the traditional workhorses, he may need all kinds of combinatorics, algebra, probability, and so on.

In addition, the applied scientist needs to cope increasingly with the nonlinear world and the extra mathematical sophistication that this requires. For that is where the rewards are. Linear models are honest and a bit sad and depressing: proportional efforts and results. It is in the non-linear world that infinitesimal inputs may result in macroscopic outputs (or vice versa). To appreciate what I am hinting at: if electronics were linear we would have no fun with transistors and computers; we would have no TV; in fact you would not be reading these lines.

There is also no safety in ignoring such outlandish things as nonstandard analysis, superspace and anticommuting integration, p-adic and ultrametric space. All three have applications in both electrical engineering and physics. Once, complex numbers were equally outlandish, but they frequently proved the shortest path between 'real' results. Similarly, the first two topics named have already provided a number of 'wormhole' paths. There is no telling where all this is leading - fortunately.

Thus the original scope of the series, which for various (sound) reasons now comprises five subseries: white (Japan), yellow (China), red (USSR), blue (Eastern Europe), and green (everything else), still applies. It has been enlarged a bit to include books treating of the tools from one subdiscipline which are used in others. Thus the series still aims at books dealing with:

- a central concept which plays an important role in several different mathematical and/or scientific specialization areas;
- new applications of the results and ideas from one area of scientific endeavour into another;
- influences which the results, problems and concepts of one field of enquiry have, and have had, on the development of another.

Convolutions and convolutional equations abound all over mathematics and its applications. Convolutional calculus, in this systematic and unifying book on the topic, is presented as the systematic search and analysis for convolutional products in the sense of the following definition: a convolution for a linear operator L (on some linear (function) space) is a product $\star$: $\mathfrak{X}\times\mathfrak{X}\rightarrow\mathfrak{X}$ such that $L(f\star g) = (Lf)\star g$ for all f,g. As such, convolutional calculus embraces the same territory as the general theory of multipliers (apart from the point of view).

Apart from their intrinsic interest, convolutions are at the basis of the construction of operational calculi and are important in the algebraic approach to analysis and the construction of right inverses to the operation of differentiation; all topics which are dealt with systematically in this volume.

The shortest path between two truths in the real domain passes through the complex domain.

J. Hadamard

La physique ne nous donne pas seulement l'occasion de résoudre des problèmes ... elle nous fait pressentir la solution.

H. Poincaré

Never lend books, for no one ever returns them; the only books I have in my library are books that other folk have lent me.

Anatole France

The function of an expert is not to be more right than other people, but to be wrong for more sophisticated reasons.

David Butler

Bussum, January 1990

Michiel Hazewinkel

CONTENTS

PREFACE

> Convolution algebras abound in both analysis and algebra. Many have been exhaustively studied ; others, of equal interest and possibly of equal importance, have hardly been noticed.
>
> *E. Hewitt* and *K. Ross*

The neologism "convolutional calculus" could only sporadically be met in mathematical writings. However, the term "convolution" is commonly used in analysis, though its meaning is far from univalent. Best known examples of convolutions are the convolution of the Fourier transform and the Duhamel convolution, the latter conceived as a special case of the first. In an abstract setting, convolutions are introduced for functions on locally compact topological groups. Broadly speaking, a convolution is always conceived as a bilinear, commutative and associative operation in a linear space. In other words, it is a "multiplication" in a linear space, such that the space itself becomes a commutative and associative algebra.

This book is an attempt at systematization of the search for new convolutions in commonly used linear spaces of analysis, and their uses. In a sense, this is a pioneer work with all the faults of such an undertaking. Nevertheless, we believe that it contains some new and useful information. The author's guidelines have been a definition of a convolution of a linear operator:

A bilinear, commutative and associative operation $*: \mathfrak{X}\times\mathfrak{X}\to\mathfrak{X}$ in a linear space $\mathfrak{X}$ is said to be a convolution of a linear operator $L:\mathfrak{X}\to\mathfrak{X}$ iff the relation $L(f*g)=(Lf)*g$ holds for all $f, g\in\mathfrak{X}$.

In the sense of this definition, the Duhamel convolution is a convolution of the Volterra integration operator. Such a definition, proposed by the author about 1966, is nothing but an inversion of the definition of a multiplier of a commutative and associative algebra. That is why the general theory of multipliers should have the same scope as the convolutional calculus. The differences should be a matter of viewpoint only. The purpose of the convolutional calculus, at least as the author perceives it, should consist in extracting new convolutions from various linear problems of analysis, in determining their multipliers, and in using these convolutions and their multipliers. As in the theory of multipliers, only annihilators-free convolutions are of any use. One of the applications of such convolutions is in developing corresponding operational calculi. The convolution quotients approach, descending from Volterra, had been used by Mikusiński as a direct algebraic basis of the Heaviside operational calculus. The brilliant approach of Mikusiński is still a pattern in the development of new operational calculi. However, as it had been noticed first by Máté, Mikusiński's approach is impracticable when all the elements of a convolutional algebra are its divisors of zero. In such cases multiplier quotients instead of convolution quotients can be used. When both approaches are applicable, they are equivalent, but in a sense the multiplier quotients approach is more natural. A linear operator may have many annihilators-free convolutions with one and the same multipliers ring.

The author is not the first to notice the relation between the Duhamel convolution and its multipliers ring. This relation had been conceived and used at least by Máté, Száz, Boehme, Struble, Krabbe and Weston. The author is possibly the first to use some notions from the theory of right inverse operators, developed by D. Przeworska-Rolewicz in convolutional calculus. At any rate, he was one of the first to seek convolutions of the right inverse operators of differentiation and its square. It may be interesting to describe briefly the hunt for convolutions of the integration operators, i. e. the right inverses of the differentiation operator. There is a common prejudice that the only linear operation, inverse to the differentiation, is the integration from a fixed point. In 1935 J. Delsarte definitely perceived that this was not the case. He began a systematic study of these generalized integration operators and their Taylor formulas. He even proposed a generalization of the Duhamel convolution, intended for such generalized integration operators. In spite of the commutativity of Delsarte's "convolution", it is not an associative one, except in the quite narrow space of mean-periodic functions. The convolution problem for the generalized integration operator was solved simultaneously by the author and by L. Berg in 1974. Except for the operational calculus, these convolutions can be used as an algebraic approach to problems of analysis. Here we give an application of such convolutions to the problem of expanding analytic functions on a given system of exponents.

The convolution problem for the most general right inverse of the square of differentiation still remains open. But there are general classes of right inverses of the square of differentiation, for which explicit convolutions have been found. By means of Delsarte-Povzner transmutation operators these convolutions can be transformed into convolutions of right inverses of linear differential operators of the second order with variable coefficients. The Sturm-Liouville problem is embraced as a special case. Thus the problem of determining explicit convolutions for the Sturm-Liouville integral transformations is solved. The relevant results of the author and N. Bozhinov are presented here. All these convolutions can be used in various problems of mathematical physics. Some new Duhamel-type representations of non-local boundary value problems for the heat and Laplace equations are obtained. Such representations can be used as an alternative to the difference methods developed ad hoc for these problems.

In this book ideas of various sources could be traced. The author acknowledges gratefully the fruitful influence of many foreign mathematicians, some of whom he has the privilege of knowing personally. Here is an incomplete list of them: J. Mikusiński, C. Ryll-Nardzewski, D. Przeworska-Rolewicz, P. Antosic, W. Kierat, L. Berg, M. Tasche, H.-J. Glaeske, E. Krätzel, V. A. Ditkin, A. P. Prudnikov, A. Y. Povzner, B. M. Levitan, A. I. Botashev, E. Gesztelyi, Á. Száz, A. Bleyer, T. K. Boehme, R. Struble, G. Krabbe, B. Stanković. Many of the ideas developed in this book arose in discussions with them. But they by no means bear any responsibility for any possible faults. The author is more than thankful to M. Tasche, who willingly took on himself the burden to read the whole raw manuscript. His comments, remarks and the numerous errors and slips discovered were an indispensible help to the author. The author thanks his collaborators and the participants in the seminar on "Operational Calculus and Linear Systems": N. Bozhinov, D. Mineff, S. Grozdev, V. Kiryakova, and R. Petrova for many suggestive discussions. Special thanks are due to the first three of them and to G. Tchobanov for their selfless help in preparing the manuscript. It is a pleasure to acknowledge the efforts made by the editors N. Chakalova and M. Pecheva to improve the manuscript. Last but not least, the author is grateful to Professor M. Todorov, to Professor I. Tchobanov and to Academician L. Iliev for their support and encouragement.

The author

PREFACE TO THE SECOND EDITION

Compared to the first Bulgarian edition this second edition appears in the *East European Series* only with minor changes. Those, who are interested in further developments of the ideas and methods of the convolutional calculus may consult the book: N. B o z h i n o v, *Convolutional representations of commutants and multipliers*. Publ. House of the Bulgarian Academy of Sciences, Sofia, 1988.

The author thanks P. Petrov and M. Vassilev for the help in debugging misprints and errors in the first edition and Dr. David J. Larner, head of Science and Technology division of the *Kluwer Academic Publishers* for the proposition the second edition of this book to be published in the *East European Series*.

Sofia, September 27, 1989

The author

CHAPTER 1

CONVOLUTIONS OF LINEAR OPERATORS. MULTIPLIERS AND MULTIPLIER QUOTIENTS

1.1. THE DUHAMEL CONVOLUTION

The *Duhamel convolution*

(1) $$(f * g)(t) = \int_0^t f(t-\tau) g(\tau)\, d\tau$$

is of considerable importance in many problems of analysis and its applications. It had been put by Mikusiński [77] as a basis of his direct approach to the Heaviside operational calculus. This approach is based on the connection between the *Volterra integration operator*

(2) $$lf(t) = \int_0^t f(\tau) d\tau$$

and the Duhamel convolution. This connection is expressed saying that l is the convolutional operator $\{1\}*$, i. e. $lf = \{1\} * f$. This relation allows to obtain explicit representations of the commutants of Volterra's integration operator in various function spaces. The considerations in the present section are a preliminary but typical illustration of the general convolutional scheme developed in the next sections.

1.1.1. Algebraic and functional properties of the Duhamel convolution. Our basic spaces in which we consider the Duhamel convolution (1) are the space $\mathscr{C}(\Delta)$ of the continuous functions in an interval Δ, containing the point 0, and the space $\mathscr{L}(\Delta)$ of the summable (or locally summable) functions in such an interval. In the case of a segment $\Delta = [\alpha, \beta]$ the topology in $\mathscr{C}(\Delta)$ is defined by means of the *uniform norm* $\| f \|_{\alpha,\beta} = \max\limits_{\alpha \le t \le \beta} | f(t)|$ and in other cases by a system of seminorms of the same kind, but with segments $[\alpha, \beta] \subset \Delta$. The corresponding convergence in the last case is said to be *almost uniform convergence.* The topology in $\mathscr{L}(\Delta)$ is defined through a system of seminorms $\|f\|_{\alpha,\beta} = \int_\alpha^\beta |f(t)|\, dt$, where $[\alpha, \beta] \subset \Delta$ and $\alpha \le 0 \le \beta$.

The Duhamel convolution (1) is an operation well defined both in $\mathscr{C}(\Delta)$ and in $\mathscr{L}(\Delta)$, i. e. $*: \mathscr{C}(\Delta)\times\mathscr{C}(\Delta) \to \mathscr{C}(\Delta)$ and $*: \mathscr{L}(\Delta)\times\mathscr{L}(\Delta)\to\mathscr{L}(\Delta)$. Obviously, it is a bilinear and commutative operation. Moreover, it is a separately continuous operation, i. e. $f_n \to f$ implies $f_n * g \to f * g$ for each g (in the corresponding space and topology).

It is well known that (1) is an associative operation in an one-sided interval Δ of 0. It is not difficult to obtain the *associativity property* of (1) for a two-sided interval Δ, but we prefer to give a direct proof of the associativity of (1) for arbitrary interval Δ containing the point $t=0$. The idea of the proof applies in more general situations too. We restrict ourselves to the space $\mathscr{C}(\Delta)$.

Proof of the associativity of $*$ in $\mathscr{C}(\Delta)$. Let λ and μ, $\lambda \neq \mu$ be real numbers. Then

$$\{e^{\lambda t}\} * \{e^{\mu t}\} = \frac{e^{\lambda t} - e^{\mu t}}{\lambda - \mu},$$

if λ, μ and ν are mutually different numbers. Then we get

$$(\{e^{\lambda t}\} * \{e^{\mu t}\}) * \{e^{\nu t}\} = \frac{e^{\lambda t}}{(\lambda-\mu)(\lambda-\nu)} + \frac{e^{\mu t}}{(\mu-\lambda)(\mu-\nu)} + \frac{e^{\nu t}}{(\nu-\lambda)(\nu-\mu)}.$$

Since this expression is completely symmetric with respect to λ, μ and ν then the relation

$$(\{e^{\lambda t}\} * \{e^{\mu t}\}) * \{e^{\nu t}\} = \{e^{\lambda t}\} * (\{e^{\mu t}\} * \{e^{\nu t}\})$$

holds. Now let us differentiate the last equation m times with respect to λ, n times with respect to μ, and p times with respect to ν, Thus we get

$$(\{t^m e^{\lambda t}\} * \{t^n e^{\mu t}\}) * \{t^p e^{\nu t}\} = \{t^m e^{\lambda t}\} * (\{t^n e^{\mu t}\} * \{t^p e^{\nu t}\}).$$

Letting $\lambda \to 0$, $\mu \to 0$ and $\nu \to 0$, we get

$$(\{t^m\} * \{t^n\}) * \{t^p\} = \{t^m\} * (\{t^n\} * \{t^p\})$$

for $m, n, p = 0, 1, 2, \ldots$. This proves the relation

$$(f * g) * h = f * (g * h)$$

for arbitrary polynomials f, g and h, due to the bilinearity of (1). Further, using the Weierstrass approximation theorem and the separate continuity of (1), the associativity relation for arbitrary real-valued f, g and h from $\mathscr{C}(\Delta)$ follows immediately. The bilinearity of (1) allows to reduce the case of complex-valued functions to the case of real-valued functions. □

The associativity of (1) in $\mathscr{L}(\Delta)$ is easily shown reducing it to the associativity in $\mathscr{C}(\Delta)$. In fact, if f, g and h are from $\mathscr{L}(\Delta)$, then lf, lg, $lh \in \mathscr{C}(\Delta)$ and hence

$$[(lf) * (lg)] * (lh) = (lf) * [(lg) * (lh)],$$

i. e. $l^3[(f * g) * h] = l^3[f * (g * h)]$. If we differentiate the last equation three times with respect to t, we get the associativity relation $(f * g) * h = f * (g * h)$ in $\mathscr{L}(\Delta)$.

Therefore, the spaces $\mathscr{C}(\Delta)$ and $\mathscr{L}(\Delta)$ with the operations $f+g=\{f(t) + g(t)\}$, $\lambda f = \{\lambda f(t)\}$ with $\lambda \in \mathbb{C}$ and the Duhamel convolution $*$ as a multiplication are commutative and associative algebras. We name them *Duhamel convolution algebras* and use the notations $\mathscr{C}(\Delta)$ and $\mathscr{L}(\Delta)$ for them instead of the more exact $[\mathscr{C}(\Delta), *]$ and $[\mathscr{L}(\Delta), *]$.

A commutative algebra $\mathscr{A}$ with multiplication $a*b$ is said to be *annihilators-free* iff there is no element $a\neq 0$ in $\mathscr{A}$, such that $a*b=0$ for each $b\in\mathscr{A}$. That is to say in an annihilators-free algebra $\mathscr{A}$ the fulfilling of the equality $a*b=0$ for each $b\in\mathscr{A}$ implies $a=0$.

Lemma 1. *The Duhamel convolution algebras $\mathscr{C}(\Delta)$ and $\mathscr{L}(\Delta)$ are annihilators-free.*

Proof. Indeed, if $f\in\mathscr{C}(\Delta)$ is such that $f*g=0$ for each $g\in\mathscr{C}(\Delta)$, then in particular $f*\{1\}=0$, e. g. $lf=0$. Hence, by differentiation we get $f=0$. □

The next question for the Duhamel convolution is the problem of exact characterization of the *divisors of zero* in the corresponding spaces. An element $f\neq 0$ of a Duhamel convolution algebra is said to be a divisor of zero iff there exists a non-zero element g of this algebra such that $f*g=0$.

In the case when $\Delta=[0,\infty)$ (or $\Delta=(-\infty, 0]$) the answer is given by an important theorem of Titchmarsh [101]: The convolutional algebras $\mathscr{C}([0,\infty])$ and $\mathscr{L}([0,\infty])$ have no divisors of zero.

It is readily seen using *Titchmarsh's theorem* that the only divisors of zero in the Duhamel convolution algebras $\mathscr{C}(-\infty,\infty)$ and $\mathscr{L}(-\infty,\infty)$ are the functions with supports on the half-lines $t\geqq 0$ or $t\leqq 0$.

In the case of one-sided finite interval $\Delta=[0, T]$ or $\Delta=[-T, 0]$ again by Titchmarsh it is proved that $f\in\mathscr{C}(\Delta)$ is a divisor of zero of (1) iff there is a number α, $0<\alpha\leqq T$, such that $f(t)=0$ for $0\leqq t\leqq\alpha$. The same statement is valid in $\mathscr{L}(\Delta)$ too, but with equality holding almost everywhere.

An elementary proof of Titchmarsh's theorems for finite or infinite interval, due to Mikusiński and Ryll-Nardzewski, can be seen in the book of Mikusiński [79]. Another proof using methods from functional analysis is given by Kalish [67].

Through combining the two Titchmarsh's theorems, a description of the divisors of zero of the Duhamel convolution algebras $\mathscr{C}(\Delta)$ can be given in the case of half-infinite or finite interval containing the zero point in its interior.

Lemma 2. *Let Δ be a segment $[\alpha,\beta]$, with $\alpha<0<\beta$. A function $f\in\mathscr{C}(\Delta)$ is a divisor of zero of (1) iff there is a one-sided vicinity of zero, such that $f(t)=0$ in it.*

Proof. If we denote $\Delta_-=[\alpha, 0]$ and $\Delta_+=[0,\beta]$, then each $f\in\mathscr{C}(\Delta)$ can be represented in the form $f=f_-+f_+$, where $f_-(t)=f(t)$ in Δ_- and $f_-(t)=0$ in $\Delta\setminus\Delta_-$; and $f_+(t)=f(t)$ in Δ_+ and $f_+(t)=0$ in $\Delta\setminus\Delta_+$. Then obviously $f*g=(f*g)_-+(f*g)_+=f_-*g_-+f_+*g_+$. If f is a function from $\mathscr{C}(\Delta)$ which vanishes in a one-sided vicinity $[\alpha_1, 0]$ or $[0,\beta_1]$ of the point zero, always a function $g\in\mathscr{C}(\Delta)$, $g\neq 0$, can be found, such that $f*g=0$. In fact, let e. g. $f(t)=0$ for $0\leqq t\leqq\beta_1<\beta$. Then we can take $g(t)=0$ for $\alpha\leqq t\leqq\beta_1$ and $g(t)=t-\beta_1$ for $\beta_1\leqq t\leqq\beta$. Conversely, let $f\neq\{0\}$ and $f*g=0$ for $g\neq\{0\}$. Then $f_-*g_-=0$ and $f_+*g_+=0$. Let us assume that $f_-\neq 0$. Then from $f_-*g_-=0$ it follows that either $g_-=0$, or there exists a one-sided vicinity of the point zero, in which $f(t)=0$. Let us consider the case $g_-=0$. Then $g_+\neq 0$. From $f_+*g_+=0$ it follows that there exists a one-sided vicinity $[0,\beta_1]$ of the point zero, in which $f_+(t)=0$ and hence $f(t)=0$. □

In a similar way it is possible to prove the following

Lemma 3. *Let Δ be a half-infinite interval containing the point 0 in its interior. A function $f \in \mathscr{C}(\Delta)$ is a divisor of zero of (1) either if $f(t)$ is equal to zero on the whole real half-line contained in Δ, or if $f(t)=0$ in a one-sided vicinity of $t=0$, lying in the other finite part of Δ.*

Later on we will need some more subtle properties of the Duhamel convolution (1). A list of such properties can be seen in Mikusiński and Ryll-Nardzewski [80].

Lemma 4. *Let Δ be an interval containing the point 0. If $f \in \mathscr{C}(\Delta)$ is a function with (locally) bounded variation, and $g \in \mathscr{C}(\Delta)$, then $f * g$ is a smooth function in Δ, i. e. $f * g \in \mathscr{C}^1(\Delta)$.*

Proof. Under the hypothesis the identity

$$f * g = l\left\{\text{RS-}\int_0^t g(t-\tau)df(\tau) + f(0)g(t)\right\}$$

is fulfilled, where by RS- it is denoted that the corresponding integral is to be understood in Riemann-Stieltjes sense. This identity can be verified easily in the case when $g \in \mathscr{C}^1(\Delta)$. If $g \in \mathscr{C}(\Delta)$, then we choose a sequence $g_n \in \mathscr{C}^1(\Delta)$ with $g_n \to g$ in the topology of $\mathscr{C}(\Delta)$, i. e. almost uniformly. Since

$$\int_0^t g_n(t-\tau)df(\tau) \to \int_0^t g(t-\tau)df(\tau)$$

in $\mathscr{C}(\Delta)$, the identity is proved. It implies $f * g \in \mathscr{C}^1(\Delta)$ and the differentiation formula

$$(f * g)'(t) = \int_0^t g(t-\tau)df(\tau) + f(0)g(t). \square \tag{3}$$

An analogous theorem for the space $\mathscr{L}(\Delta)$ follows from a general theorem of Bozhinov [11].

Lemma 5. *If a function $f \in \mathscr{L}(\Delta)$ is equal almost everywhere in Δ to a function with (locally) bounded variation, and if $g \in \mathscr{L}(\Delta)$, then $f * g$ is absolutely continuous function.*

Proof. Without loss of generality, we may assume that Δ is a right-hand side interval of 0, and that $f(t)$ is right-continuous for each $t>0$ in Δ, i. e. $f(t+0)=f(t)$ for $t>0$. Then for $g \in \mathscr{C}^1(\Delta)$ the identity

$$f * g = l\left\{\text{RS-}\int_0^t g(t-\tau)df(\tau) + f(0)g(t)\right\}$$

is fulfilled. If by μ_f we denote the Lebesgue measure on the line, generated by a right-continuous function f with (locally) bounded variation (for the construction, see e. g. Edwards [55], sects. 4.5 and 4.7), then the identity

$$\text{RS-}\int_0^t g(t-\tau)df(\tau) = \int_0^t g(t-\tau)d\mu_f = M_f g$$

is fulfilled. The operator M_f with a fixed f is a continuous operator from $\mathscr{L}(\Delta)$ into $\mathscr{L}(\Delta)$. Then we take a sequence $g_n \in \mathscr{C}^1(\Delta)$, with $g_n \to g$ in $\mathscr{L}(\Delta)$. Then, $lM_f g_n + f(0) l g_n = f * g_n$. If $n \to \infty$, we get

$$f * g = lM_f g + f(0) l g.$$

Thus we have proved that $f * g$ is absolutely continuous. For its derivative we get

$$(f * g)'(t) = \int_0^t g(t-\tau) d\mu_f + f(0) g(t). \quad \square \tag{4}$$

We should not forget that the last formula is derived under the assumption that $f(t)$ is normed with the requirement it to be *right-continuous* in the interior points of Δ. The interval Δ itself is supposed to be right-sided to $t=0$. For a two-sided interval Δ for the point $t=0$, the same formula (4) is true under the assumption that $f(t)$ is right-continuous for interior $t>0$, and left-continuous for interior $t<0$.

1.1.2. Multipliers of the Duhamel convolution in $\mathscr{C}(\Delta)$. Let us consider the Duhamel convolution algebra $\mathscr{C}(\Delta)$. It is connected with the Volterra integration operator l. In the previous section we have used the identity

$$l(f * g) = (lf) * g,$$

which is an immediate result of the associativity of the Duhamel convolution, due to $lf = \{1\} * f$.

An important problem connected with the Duhamel convolution is to find all linear operators $M: \mathscr{C}(\Delta) \to \mathscr{C}(\Delta)$ which fulfill the condition

$$M(f * g) = (Mf) * g \tag{5}$$

for all $f, g \in \mathscr{C}(\Delta)$.

Each such operator is said to be a *multiplier* of the Duhamel convolution algebra $\mathscr{C}(\Delta)$. Sometimes we say that M is a multiplier of the convolution $f * g$. For the general multiplier theory see Larsen [72].

Lemma 6 (Larsen [72], p. 13). *Each multiplier of the Duhamel convolution algebra $\mathscr{C}(\Delta)$ is a linear and continuous operator in $\mathscr{C}(\Delta)$.*

Proof. Let $M: \mathscr{C}(\Delta) \to \mathscr{C}(\Delta)$ be a multiplier of $f * g$ in $\mathscr{C}(\Delta)$. Let f, g, h be arbitrary functions from $\mathscr{C}(\Delta)$ and let α, β be arbitrary numbers. First, from (5) we get

$$[M(\alpha f + \beta g)] * h = (\alpha f + \beta g) * (Mh) = \alpha(f * Mh) + \beta(g * Mh).$$

On the other hand,

$$(\alpha Mf + \beta Mg) * h = (\alpha Mf) * h + (\beta Mg) * h$$
$$= \alpha(Mf * h) + \beta(Mg * h) = \alpha(f * Mh) + \beta(g * Mh).$$

Thus we have

$$[M(\alpha f + \beta g) - (\alpha Mf + \beta Mg)] * h = 0.$$

But the Duhamel convolution is annihilators-free (Lemma 1) and hence

$$M(\alpha f + \beta g) = \alpha Mf + \beta Mg,$$

i. e. M is linear.

In order to establish the continuity of M with respect to the almost uniform convergence, we should use the *closed graph theorem*. Its applicability is ensured from the fact that $\mathscr{C}(\Delta)$ is a Banach or. Fréchet space. Using this theorem it is enough to prove that from $f_n \to f$ and from $Mf_n \to h$ it follows $h = Mf$. Taking an arbitrary $g \in \mathscr{C}(\Delta)$, we have $f_n * g \to f * g$ and $(Mf_n) * g \to h * g$ since (1) is a separately continuous operation. But $(Mf_n) * g = f_n * Mg \to f * (Mg) = (Mf) * g$ and hence $h * g = (Mf) * g$.

Thus we have proved $(h - Mf) * g = 0$ for all $g \in \mathscr{C}(\Delta)$. Hence, according to Lemma 1 we get $h = Mf$. Therefore, the operator M is continuous in $\mathscr{C}(\Delta)$. □

Lemma 7 (Máté [75]). *The multipliers of the Duhamel algebra $\mathscr{C}(\Delta)$ form a commutative ring.*

Proof. Let M and N be multipliers of (1), and let f, g be arbitrary functions of $\mathscr{C}(\Delta)$. Then using the characteristic property (5) of the muttipliers M and N, we get

$$(MNf) * g = M(Nf * g) = M(f * Ng)$$

and

$$(NMf) * g = Mf * Ng = M(f * Ng).$$

Hence $(MNf - NMf) * g = 0$ for $g \in \mathscr{C}(\Delta)$. From Lemma 1 we get $MNf = NMf$ for each $f \in \mathscr{C}(\Delta)$. It is almost evident that the multipliers of $\mathscr{C}(\Delta)$ form a ring. □

Now we can characterize explicitly the *multipliers of the Duhamel algebra* $\mathscr{C}(\Delta)$.

Theorem 1.1.1. *An operator $M : \mathscr{C}(\Delta) \to \mathscr{C}(\Delta)$ is a multiplier of the Duhamel convolution algebra $\mathscr{C}(\Delta)$ iff it admits an integral representation of the form*

$$Mf(t) = \frac{d}{dt} \int_0^t f(t - \tau) m(\tau) d\tau \tag{6}$$

with $m(t) \overset{\text{def}}{=} M\{1\}$ being a continuous function with (locally) bounded variation in Δ.

The function $m(t)$ is a function with bounded variation on Δ (not only locally) in the case when Δ is a finite segment.

Proof. Let $M: \mathscr{C}(\Delta) \to \mathscr{C}(\Delta)$ be a multiplier of the Duhamel convolution algebra $\mathscr{C}(\Delta)$. Since the Volterra integration operator l is a multiplier too, then $Ml = lM$ from the previous Lemma 7. If we apply the operator M to the identity $lf = \{1\} * f$, we get $Mlf = (M\{1\}) * f$, i. e. $lMf = m * f$, where $m = M\{1\} \in \mathscr{C}(\Delta)$. **Without loss of generality, we may assume $m(0) = 0$.**

Representation (6) follows immediately by differentiation. It remains to prove that m is a function with bounded variation in each finite subsegment of Δ. We consider at first the case of one-sided interval Δ with respect to the point 0. Let Δ be an interval of the form $[0, T]$, $[0, T)$ or $[0, \infty)$. We fix an arbitrary $t \in \Delta$, $t > 0$ and consider $Mf(t)$ as a linear functional on the space $\mathscr{C}([0, t])$. This is justified from the representation (6). Further it is seen that the values of $Mf(t)$ for $t \in [0, T]$ use only values of $f(\tau)$ for $\tau \in [0, t]$. According to the well-known Riesz representation theorem in $\mathscr{C}([0, t])$ we can write

$$Mf(t) = \int_0^t f(\tau)da_t(\tau),$$

where $a_t(\tau)$ is a function with bounded variation on $[0, t]$. This function is uniquely determined if we suppose it continuous from right, i. e. $a_t(\tau+0) = a_t(\tau)$ for $0 \leq \tau < t$, and normed with the condition $a_t(t) = m(0)$. We shall show that $a_t(\tau) = m(t-\tau)$ for $0 < \tau \leq t$. Indeed, using the commutation relation $Ml = lM$, we have

$$lMf(t) = \int_0^t (lf)(\tau)da_t(\tau) = \int_0^t f(\tau)d_\tau\left\{\int_\tau^t a_t(\sigma)\,d\sigma\right\},$$

but from (6) we get

$$lMf(t) = -\int_0^t f(\tau)d_\tau\left\{\int_0^{t-\tau} m(\sigma)\,d\sigma\right\}.$$

Looking on the last two identities as on Riesz representations of the linear functional $lMf(t)$ on $\mathscr{C}([0, t])$, then from the uniqueness theorem of the Riesz representation we get

$$\int_\tau^t a_t(\sigma)d\sigma = -\int_0^{t-\tau} m(\sigma)d\sigma, \quad 0 < \tau \leq t.$$

Therefore, by differentiation on τ we obtain

$$a_t(\tau) = m(t-\tau), \quad 0 < \tau \leq t. \tag{7}$$

Hence $m(\tau)$ is a function with bounded variation on $[0, t]$ since $a_t(\tau)$ is such a function. Thus the one-sided case is settled.

The case of a two-sided interval Δ with respect to the point $t=0$ can be reduced easily to the case of one-sided interval. As in the proof of Lemma 2, we may write $f = f_- + f_+$ and $m = m_- + m_+$. Then representation (6) can be written in the form $Mf = M_-f_- + M_+f_+$, where $M_-f_- = (m_- * f_-)'$ and $M_+f_+ = (m_+ * f_+)'$ are operators commuting with l in $\mathscr{C}(\Delta_-)$ and $\mathscr{C}(\Delta_+)$, correspondingly. Then, using the statement for a one-sided interval, we can assert that m_- and m_+ are functions with bounded variation in each subsegment of Δ_- and Δ_+, correspondingly. Then $m = m_- + m_+$ shall be a function with bounded variation in each finite subsegment of Δ.

Conversely, if m is a function from $\mathscr{C}(\Delta)$ with bounded variation on each subsegment of Δ, then expression (6) defines an operator $M\colon \mathscr{C}(\Delta) \to \mathscr{C}(\Delta)$, according to Lemma 4. It can be verified directly that M is in fact a multiplier of the Duhamel convolution $f * g$. □

In Dimovski [39] and in Bozhinov and Dimovski [20] this theorem is proved in another way, using a theorem of Weston [104] and [105].

Corollary. The correspondence $M \mapsto m$ between the multipliers of the Duhamel convolution algebra $\mathscr{C}(\Delta)$ and the continuous functions with locally bounded variation on Δ is an isomorphism between the algebra of these multipliers and the algebra of the functions from $\mathscr{C}(\Delta)$ with locally bounded variation on Δ with the multiplication

$$(8)\qquad (m \tilde{*} n)(t)=\frac{d}{dt}\int_0^t m(t-\tau)n(\tau)d\tau.$$

Proof. According to the theorem, this correspondence is one-to-one. Evidently, it is an isomorphism of the corresponding linear spaces. If M and N are multipliers of the Duhamel convolution algebra $\mathscr{C}(\Delta)$ with representations (6) of the form $Mf=(m*f)'$ and $Nf=(n*f)'$ correspondingly, then it is, easy to see that operator $P=MNf$ has a representation $Pf=(p*f)'$ of form (6) with $p=(m*n)'=m\tilde{*}n$. The fact that $p=(m*n)'$ is a function from $\mathscr{C}(\Delta)$ follows from Lemma 4. The fact that $(m*n)'$ is a function with locally bounded variation is subtler and it can be shown either using representation (6) for P, or referring to a more general theorem proved in Bozhinov [11]. □

In such a way, we have a complete characterization of the multipliers of the Duhamel convolution algebra $\mathscr{C}(\Delta)$.

Similar results are valid for the Duhamel convolution algebra $\mathscr{L}(\Delta)$ too, but in order to establish them we shall use some results which will be proved in the next section.

1.1.3. Duhamel representations of the commutants of the Volterra integration operator. We have already seen (Lemma 7) that the multipliers of the Duhamel convolution in $\mathscr{C}(\Delta)$ are linear continuous operators, which commute with the Volterra integration operator l. (Just in the same way this can be seen for $\mathscr{L}(\Delta)$.) Now we shall show that the converse is also true.

Theorem 1.1.2. *A linear continuous operator $M:\mathscr{C}(\Delta)\to\mathscr{C}(\Delta)$ (or $M:\mathscr{L}(\Delta)\to\mathscr{L}(\Delta)$) commutes with the Volterra integration operator l in $\mathscr{C}(\Delta)$ (or in $\mathscr{L}(\Delta)$) iff it is a multiplier of the corresponding Duhamel convolution algebra.*

Proof. First we shall prove this for the Duhamel convolution subalgebra of the polynomials. Then we shall use a pproximation by polynomials in order to prove the assertion for the whole convolutional algebras $\mathscr{C}(\Delta)$ or $\mathscr{L}(\Delta)$. Having in mind Lemma 7, only the first part of the statement remains to be proved. Let M: $\mathscr{C}(\Delta)\to\mathscr{C}(\Delta)$ be a continuous linear operator which commutes with l. If m and n are arbitrary non-negative integers, then we apply the operator l^{m+n} to the evident identity $(M\{1\})*\{1\}=\{1\}*(M\{1\})$. Using the commuting of l with the Duhamel convolution and with the operator M, we get $Ml^m\{1\}*l^n\{1\}=l^m\{1\}*Ml^n\{1\}$. But $l^n\{1\}=t^n/n!$ and $l^m\{1\}=t^m/m!$, and thus we have proved

$$M\{t^m\}*t^n=t^m*M\{t^n\}$$

for $m, n=0, 1, 2,\dots$. From these identities and from the bilinearity of the Duhamel convolution, it follows that the relation $Mf*g=f*Mg$ is satisfied for polynomials f and g. Now, using Weierstrass' approximation theorem, we can easily prove that

$$(9)\qquad (Mf)*g=f*(Mg)\quad \text{for all}\quad f, g\in\mathscr{C}(\Delta).$$

Now it is not difficult to establish the full multiplier relation $M(f*g)=(Mf)*g$.

To this aim, let us take an arbitrary function $h \in \mathscr{C}(\Delta)$. Using only relation (9) and the associativity of the Duhamel convolution, we get

$$M(f*g)*h=(f*g)*Mh$$

and

$$[(Mf)*g]*h=f*[(Mg)*h]=f*(g*Mh),$$

i. e. $[M(f*g)]*h=[(Mf)*g]*h$ for each $h \in \mathscr{C}(\Delta)$. Then, from Lemma 1 it follows that $M(f*g)=(Mf)*g$. Hence, M is a multiplier of the Duhamel convolution algebra $\mathscr{C}(\Delta)$.

The considerations for the Duhamel convolution algebra $\mathscr{L}(\Delta)$ proceed in a similar way. □

If we combine Theorem 1.1.2 with Theorem 1.1.1 we get the following

Corollary. A linear continuous operator $M:\mathscr{C}(\Delta)\to\mathscr{C}(\Delta)$ commutes with the Volterra integration operator l in $\mathscr{C}(\Delta)$ iff: (a) $m(t)\overset{\text{def}}{=}M\{1\}\in\mathscr{C}(\Delta)$ is a function with locally bounded variation on Δ; (b) the operator M has a representation of the form $Mf=(m*f)'$ for $f\in\mathscr{C}(\Delta)$.

Now we shall find a similar representation of the commutant of l in $\mathscr{L}(\Delta)$. According to Theorem 1.1.2 and using the same argument as in the proof of Theorem 1.1.1, we can establish a representation of the form (6) too. It remains only to characterize the function $m(t)\overset{\text{def}}{=}M\{1\}$. Dixmier [51] and Edwards [54] had found that the function $m(t)\in\mathscr{L}(\Delta)$ is always equal almost everywhere to a function with locally bounded variation in Δ. The proof is rather involved, and we drop it, referring the reader to the original papers of Dixmier and Edwards. Conversely, if $m(t)$ is equal almost everywhere to a function with locally bounded variation on Δ, from Lemma 5 it follows that the operator $Mf=(m*f)'$ is well defined in $\mathscr{L}(\Delta)$. Using the identities established in the proof of the same lemma, it can easily be seen that $Ml=lM$. This allows us to state the Dixmier-Edwards theorem in the following form:

Theorem 1.1.3. *A linear continuous operator $M:\mathscr{L}(\Delta)\to\mathscr{L}(\Delta)$ commutes with the Volterra integration operator iff*: (a) $m\overset{\text{def}}{=}M\{1\}$ *is equal almost everywhere in Δ to a function with locally bounded variation,* (b) *the operator M admits a representation of the form $Mf=(m*f)'$ for $f\in\mathscr{L}(\Delta)$.*

The commutation relation $Mlf(t)=lMf(t)$ in $\mathscr{L}(\Delta)$ is to be verified first for $f\in\mathscr{C}^1(\Delta)$, and then an arbitrary $f\in\mathscr{L}(\Delta)$ to be approximated by functions from $\mathscr{C}^1(\Delta)$, but in the topology of $\mathscr{L}(\Delta)$.

Corollary. An operator $M:\mathscr{L}(\Delta)\to\mathscr{L}(\Delta)$ is a multiplier of the Duhamel convolution algebra $\mathscr{L}(\Delta)$ iff it admits a representation of the form $Mf=(m*f)'$, where $m\overset{\text{def}}{=}M\{1\}$ is equal almost everywhere in Δ to a function with locally bounded variation.

This corollary which is a complete analogon of Theorem 1.1.1 is an immediate consequence of Theorems 1.1.2 and 1.1.3.

1.1.4. Representations of all possible continuous convolutions of the Volterra integration operator. A bilinear, commutative and associative operation $\tilde{*}$ in $\mathscr{C}(\Delta)$ or $\mathscr{L}(\Delta)$ is said to be a convolution of l iff l is a multiplier of the corresponding algebra $[\mathscr{C}(\Delta), \tilde{*}]$ or $[\mathscr{L}(\Delta), \tilde{*}]$.

The same definition can be stated so: a bilinear, commutative and associative operation $\tilde{*}$ is said to be a *convolution of l*, iff

$$l(f\tilde{*}g)=(lf)\tilde{*}g \text{ for each } f \text{ and } g. \tag{10}$$

The results established in the previous two sections allow a fair complete description of the possible separately continuous convolutions of l in $\mathscr{C}(\Delta)$ and $\mathscr{L}(\Delta)$.

A convolution $\tilde{*}$ in $\mathscr{C}(\Delta)$ is said to be separately continuous in $\mathscr{C}(\Delta)$ iff $f_n\to f$ implies $f_n\tilde{*}g\to f\tilde{*}g$ for each $g\in\mathscr{C}(\Delta)$. The same definition for a separately continuous convolution can be given for $\mathscr{L}(\Delta)$ too.

Theorem 1.1.4. *If $\tilde{*}$ is a separately continuous convolution of l in $\mathscr{C}(\Delta)$, then the representation*

$$f\tilde{*}g=\frac{d^2}{dt^2}\,[\varrho*(f*g)] \tag{11}$$

holds with $\varrho=\{1\}\tilde{*}\{1\}$. Conversely, if $\varrho\in\mathscr{C}(\Delta)$ is a function such that the expression in the right-hand side of (11) *exists in $\mathscr{C}(\Delta)$ for all $f, g\in\mathscr{C}(\Delta)$, then* (11) *is a separately continuous convolution of l in $\mathscr{C}(\Delta)$.*

Proof. Let us look on $f\tilde{*}g$ as on an operator M_g applied on f, i. e. $M_gf=f\tilde{*}g$. Then the operator M_g is linear, continuous, and it commutes with l in $\mathscr{C}(\Delta)$. In fact, $lM_gf=l(f\tilde{*}g)=(lf)\tilde{*}g=M_glf$, according to (10). The corollary of Theorem 1.1.2 gives the representation

$$M_gf=\frac{d}{dt}\,[(M_g\{1\})*f].$$

We can look on the continuous function with bounded variation $M_g\{1\}=M_{\{1\}}g$ again as on an operator applied on g. The linear continuous operator $M_{\{1\}}$ commutes with l too and hence admits a representation of the form

$$M_{\{1\}}g=\frac{d}{dt}\,[(\{1\}\tilde{*}\{1\})*g].$$

Thus we get

$$f\tilde{*}g=\frac{d}{dt}\left[\frac{d}{dt}(\varrho*g)*f\right]=\frac{d^2}{dt^2}\,(\varrho*(f*g)),$$

and the representation (11) is proved. Here $\varrho=1\tilde{*}1$ must be not only continuous, but with bounded variation too.

Let now $\varrho\in\mathscr{C}(\Delta)$ be a function such that the expression $(d^2/dt^2)(\varrho*f*g)$ exists as a function from $\mathscr{C}(\Delta)$. We shall show that the operation $\tilde{*}$, defined by (11), is a separately continuous convolution of l in $\mathscr{C}(\Delta)$.

For a proof we use the closed graph theorem. Let $f_n\to f$ and let us assume that $f_n\tilde{*}g\to h$. Then $l^2(f_n\tilde{*}g)=\varrho*f_n*g\to\varrho*f*g=l^2(f\tilde{*}g)$. Therefore $l^2h=l^2(f\tilde{*}g)$ and hence $h=f\tilde{*}g$.

The associativity relation $(f\tilde{*}g)\tilde{*}h=f\tilde{*}(g\tilde{*}h)$ can be proved easier for functions from $\mathscr{C}^1(\Delta)$. Then $f=f(0)+lf'$, $g=g(0)+lg'$ and $h=h(0)+lh'$. If we substitute f, g and h in the associativity relation, then using the properties of the Duhamel convolution it is not difficult to verify the associativity in $\mathscr{C}^1(\Delta)$. Then we can use approximation with smooth functions in order to establish the associativity in $\mathscr{C}(\Delta)$. □

The convolutional property (10) can be verified directly.

A convolution of the form (11) can have *annihilators*. Such is the case when ϱ is a divisor of zero of the Duhamel convolution.

An exact analogon of Theorem 1.1.4 can be proved for the space $\mathscr{L}(\Delta)$ too, but with ϱ equal almost everywhere to a function with (locally) bounded variation on Δ. In $\mathscr{L}(\Delta)$ a stronger result can be proved. To this aim we need the following

Lemma 8. If $\tilde{*}$ *is a separately continuous convolution of* l *in* $\mathscr{L}(\Delta)$, *then for every* f, g, $h\in\mathscr{L}(\Delta)$ *the identity*

(12) $$f\tilde{*}(g*h)=(f\tilde{*}g)*h$$

holds.

Proof. Let us consider the operator $M_fg=f\tilde{*}g$ in $\mathscr{L}(\Delta)$. From the hypothesis that $f\tilde{*}g$ is a convolution of l in $\mathscr{L}(\Delta)$ it follows that l commutes with M_f in $\mathscr{L}(\Delta)$. Therefore, according to Theorem 1.1.2, M_f is a multiplier of the Duhamel convolution, i. e. $M_f(g*h)=(M_fg)*h$ for all g, $h\in\mathscr{L}(\Delta)$. But this is relation (12). □

Theorem 1.1.5. *A separately continuous operation* $\tilde{*}$ *in* $\mathscr{L}(\Delta)$ *is a convolution of* l *in* $\mathscr{L}(\Delta)$ *iff it admits a representation of the form*

(13) $$f\tilde{*}g=\frac{d}{dt}[r*(f*g)]$$

with a function $r\in\mathscr{L}(\Delta)$, *equal almost everywhere to a function with bounded variation in* Δ.

Proof. Let $\tilde{*}$ be a separately continuous convolution of l in $\mathscr{L}(\Delta)$. In the same way as in the proof of Theorem 1.1.4 it can be seen that representation of the form (11) with $\varrho\in\mathscr{L}(\Delta)$ holds. Let us consider the linear operator

(14) $$Mf=\frac{d^2}{dt^2}(\varrho*f).$$

It is defined for each $f\in\mathscr{L}(\Delta)$, which can be factorized in the form $f=f_1*f_2$ with f_1, $f_2\in\mathscr{L}(\Delta)$. But according to a theorem due to Cohen [26] for factorizing in Banach algebras and an extension of it due to Boehme [7], each function $f\in\mathscr{L}(\Delta)$ can be factorized in this way. Therefore the operator M is defined in the whole space $\mathscr{L}(\Delta)$ and maps $\mathscr{L}(\Delta)$ into itself. We shall prove that M is a multiplier of the Duhamel convolution. Indeed, if $f=f_1*f_2$ is a factorization of f in $\mathscr{L}(\Delta)$ and $g\in\mathscr{L}(\Delta)$, then $M(f*g)=f\tilde{*}g$ $=(f_1*f_2)\tilde{*}g=(f_1\tilde{*}f_2)*g=Mf*g$, according to Lemma 8. Then from the corollary of Theorem 1.1.3 the operator M can be represented in the form $Mf=(r*f)'$, where $r=M\{1\}=d\varrho/dt$ is a function from $\mathscr{L}(\Delta)$, equal almost everywhere to a function with bounded variation on Δ.

Hence

$$f\tilde{*}g=M(f*g)=\frac{d}{dt}(r*(f*g))$$

and the necessity is proved. The sufficiency follows from Lemma 5. □

Thus we obtain a complete description of the possible separately continuous convolutions of l in $\mathscr{L}(\Delta)$. Theorems 1.1.4 and 1.1.5 could be obtained by a specialization of more general theorems proved in [36].

Corollary 1. A separately continuous convolution $\tilde{*}$ of the Volterra integration operator in $\mathscr{L}(\Delta)$ is an annihilators-free convolution iff the function $r(t)=(d/dt)\{1\tilde{*}1\}$ is a non-zero non-divisor of the Duhamel convolution.

Proof. We use representation (12). If r is a divisor of zero of the Duhamel convolution $*$, it can easily be seen that $\tilde{*}$ has non-zero annihilators. Indeed, let $r*\varphi=0$ for a $\varphi\in\mathscr{L}(\Delta)$, $\varphi\neq0$. Then φ is an annihilator of $\tilde{*}$ since $\varphi\tilde{*}g=0$ for each $g\in\mathscr{L}(\Delta)$. Let now r be a non-zero non-divisor of zero of the Duhamel convolution. If $\varphi\in\mathscr{L}(\Delta)$ with $\varphi\tilde{*}g=0$ for each $g\in\mathscr{L}(\Delta)$, then if we take $g=\{1\}$ we get $r*\varphi=0$. But $r\neq0$ is a non-divisor of zero and hence $\varphi=0$. This shows that $f\tilde{*}g$ is an annihilators-free convolution in $\mathscr{L}(\Delta)$. □

In the cases $\Delta=(-\infty, 0]$ or $\Delta=[0, \infty)$ the assertion can be strengthened. We confine ourselves only to the case $\Delta=[0,\infty)$.

Corollary 2. Let $\tilde{*}$ be a separately continuous non-trivial convolution of l in $\mathscr{L}(\Delta)$. Then $\tilde{*}$ has no divisors of zero.

Proof. The non-triviality of $f\tilde{*}g$ is equivalent to the requirement the function r in (13) to be a non-zero function. Let $f\tilde{*}g=0$. By applying l to this identity we get $r*(f*g)=0$. But according to the Titchmarsh theorem for the infinite interval $[0, \infty)$, it follows that $f*g=0$ and therefore either $f=0$ or $g=0$. Hence the convolution has no divisors of zero. □

Later on we shall consider only annihilators-free convolutions $\tilde{*}$ of l in $\mathscr{C}(\Delta)$ and $\mathscr{L}(\Delta)$. Of special interest for us shall be the set of all non-divisors of zero of such a convolution $\tilde{*}$. We shall denote them by $\mathscr{C}_{\tilde{*}}(\Delta)$ or $\mathscr{L}_{\tilde{*}}(\Delta)$. It happens that these sets are the same for all annihilators-free separately continuous convolutions of l on $\mathscr{C}(\Delta)$ or $\mathscr{L}(\Delta)$.

Theorem 1.1.6. *The set of non-divisors of zero of an arbitrary separately continuous and annihilators-free convolution $\tilde{*}$ of the Volterra integration operator l in $\mathscr{C}(\Delta)$ (or $\mathscr{L}(\Delta)$) coincides with the set of non-divisors of zero of the Duhamel convolution $*$ in $\mathscr{C}(\Delta)$ (or $\mathscr{L}(\Delta)$).*

Proof. Since the statement can be proved with the same argument in $\mathscr{C}(\Delta)$ and in $\mathscr{L}(\Delta)$, we shall restrict ourselves to $\mathscr{C}(\Delta)$. We use representation (11) of $f\tilde{*}g$. From this it follows immediately that if $f*g=0$, then $f\tilde{*}g=0$. Let now $f\tilde{*}g=0$. Since the operation $\tilde{*}$ is a convolution of l, then $l^2(f\tilde{*}g)=(lf)*(lg)$ and hence $(d/dt)[\varrho*(lf)*(lg)]=0$. Therefore, $\varrho*(f*g)=0$. We have shown that the requirement ϱ to be a non-divisor of zero of the Duhamel convolution is equivalent to the requirement the convolution $\tilde{*}$ to be annihilators-free. Hence $f*g=0$ and the theorem is proved. □

Open problem. Which is the exact class of the functions $\varrho \in \mathscr{C}(\Delta)$ for which representation (11) is defined in the whole space $\mathscr{C}(\Delta)$ and it is a convolution of l in $\mathscr{C}(\Delta)$? Conjecture: ϱ has the form $\varrho = lr$, with r from $\mathscr{C}(\Delta)$ and with (locally) bounded variation.

1.2. THE MIKUSIŃSKI RING

The idea to introduce quotients with respect to the Duhamel convolution is due to Volterra and Pérès [103]. The systematic development of the Heaviside operational calculus on such convolutional quotients has been done by Mikusiński [77] and [78] for the cases $\Delta=[0, \infty)$ and for finite intervals $\Delta=[0, T]$ and $\Delta=[0, T)$. The cases of $\Delta=(-\infty, +\infty)$ and of a finite interval $\Delta=(\alpha, \beta)$ with $\alpha<0<\beta$ have been treated by Krabbe [70] and Száz [93]. Instead of convolutional quotients one can use multiplier quotients, cf. Dimovski [40] and Száz [94]. The multiplier quotients approach has some advantages in developing operational calculi for functions of several variables. Since the corresponding algebraic structures are well known, here we restrict ourselves with a short survey.

1.2.1. Convolution quotients. Now we shall consider a quite natural algebraic extension of the Duhamel convolution algebra $\mathscr{C}(\Delta)$. There is no need to consider separately the Duhamel convolution algebra $\mathscr{L}(\Delta)$, since in the last case all proceeds in the same way. Moreover, the considerations for $\mathscr{L}(\Delta)$ can be reduced easily to considerations in $\mathscr{C}(\Delta)$.

We denote by $\mathscr{C}_*(\Delta)$ the set of the non-trivial non-divisors of zero of the Duhamel convolution (1) in $\mathscr{C}(\Delta)$. In considerations in which the specifying of the interval Δ is not essential we shall use the shorter notations $\mathscr{C}=\mathscr{C}(\Delta)$ and $\mathscr{C}_*=\mathscr{C}_*(\Delta)$.

According to Titchmarsh's theorem for $\Delta=[0, \infty)$, the set $\mathscr{C}_*$ consists of all functions of $\mathscr{C}$ with the exception of the zero function $\{0\}$. In the other cases Lemmas 2 and 3 give an effective characterization of $\mathscr{C}_*$. Evidently, the set $\mathscr{C}_*$ is closed under the Duhamel convolution, i. e. if $f, g \in \mathscr{C}_*$, then $f*g \in \mathscr{C}_*$.

The convolution quotients of $\mathscr{C}=\mathscr{C}(\Delta)$ are introduced by means of the factorization of the set $\mathscr{C}\times\mathscr{C}_*$ with respect to the equivalence relation

$$(f, g) \sim (h, k) \Leftrightarrow f*k = g*h. \tag{1}$$

Definition. The ring $\mathrm{M}(\Delta)=(\mathscr{C}\times\mathscr{C}_*)/_\sim$ is said to be the *Mikusiński ring* for the interval Δ.

The elements of the ring $\mathrm{M}(\Delta)$ can be considered as convolution quotients f/g, $f\in\mathscr{C}$, $g\in\mathscr{C}_*$. They may be interpreted as equivalence classes under relation (1).

The basic algebraic operations in $\mathrm{M}(\Delta)$ are defined by the equalities

$$\frac{f}{g}+\frac{h}{k}=\frac{f*k+g*h}{g*k} \tag{2}$$

and

$$\frac{f}{g}\,\frac{h}{k}=\frac{f*h}{g*k}. \tag{3}$$

It is easily seen that the Duhamel ring $\mathscr{C}(\Delta)$ can be embedded in $\mathrm{M}(\Delta)$ by the map

$$f \to \frac{lf}{\{1\}}. \tag{4}$$

That is why we can look on the elements of $\mathscr{C}(\Delta)$ as on elements of the Mikusiński ring $\mathsf{M}(\Delta)$ and write $f * g = fg$.

Moreover, it is easily seen that the ring C of the complex numbers can also be embedded in $\mathsf{M}(\Delta)$ by the map

$$\alpha \to \frac{\{\alpha\}}{\{1\}}. \tag{5}$$

Moreover, if we use again the denotation $\alpha = \{\alpha\}/\{1\}$, we see at once that if $f \in \mathscr{C}(\Delta)$, then $\alpha f = \{\alpha f(t)\}$.

Thus (1), (2) and (3) are the defining relations of the Mikusiński ring $\mathsf{M}(\Delta)$, and (4) and (5) are identifications of some elements of $\mathsf{M}(\Delta)$ as functions or as complex numbers. For the purposes of the operational calculus it is important some other elements of $\mathsf{M}(\Delta)$ to be identified as functions from $\mathscr{C}(\Delta)$.

As for the unit 1 of the Mikusiński ring, it cannot be interpreted as function, but only as the number 1. The constant function $\{1\}$ can be interpreted as the multiplier l of the Duhamel convolution $f * g$, due to the identity $lf = \{1\} * f$. The element

$$s = \frac{1}{l} = \frac{1}{\{1\}}, \tag{6}$$

which is the algebraic inverse of l, can be considered as an operator for "algebraic differentiation." Its connection with the usual differentiation operator $D = d/dt$ becomes clear from the identity

$$lf'(t) = f(t) - f(0)$$

valid in $\mathscr{C}^1(\Delta)$. From it we easily obtain

$$sf = f' + f(0), \tag{7}$$

where $f(0)$ denotes not the constant function $\{f(0)\}$, but the number $f(0)$.

If $f \in \mathscr{C}^{(n)}(\Delta)$, then from the Taylor formula

$$f(t) = f(0) + f'(0)\frac{t}{1!} + \cdots + f^{(n-1)}(0)\frac{t^{n-1}}{(n-1)!} + \frac{1}{(n-1)!}\int_0^t (t-\tau)^{n-1} f^{(n)}(\tau)\,d\tau$$

we get

$$s^n f = f^{(n)} + f^{(n-1)}(0) + f^{(n-2)}(0)s + \cdots + f(0)s^{n-1}. \tag{8}$$

The same formula could be obtained, of course, by repeated use of (7).

Lemma 1. *If* $P(\lambda) = a_0\lambda^n + a_1\lambda^{n-1} + \cdots + a_{n-1}\lambda + a_n$, $a_0 \neq 0$, *is a polynomial with complex coefficients, then the element* $P(s)$ *of the Mikusiński ring* $\mathsf{M}(\Delta)$ *is not a divisor of zero in* $\mathsf{M}(\Delta)$.

Proof. It is sufficient to prove that $s - \alpha$ is not a divisor of zero in $\mathsf{M}(\Delta)$ for all $\alpha \in \mathsf{C}$. Indeed, let $(s-\alpha)q = 0$ for some $q \in \mathsf{M}(\Delta)$. If $q = x/y$, with $x \in \mathscr{C}$ and $y \in \mathscr{C}_*$, then the above equality is equivalent to $(s-\alpha)x = 0$. By multiplying by l, we get the Volterra integral equation

$$x(t)-\alpha\int_0^t x(\tau)\,d\tau=0.$$

The only solution of this integral equation is the zero function. □

Lemma 2. *If α is an arbitrary complex number and n is a natural number, then*

$$\frac{1}{(s-\alpha)^n}=\left\{\frac{t^{n-1}}{(n-1)!}e^{\alpha t}\right\},\quad n=1,2,\ldots \tag{9}$$

Proof. By induction. From (7) we have

$$s\{e^{\alpha t}\}=\{\alpha e^{\alpha t}\}+1,$$

which proves (9) for $n=1$.

If (9) is established for some n, then from

$$\frac{1}{(s-\alpha)^{n+1}}=\frac{1}{s-\alpha}\,\frac{1}{(s-\alpha)^n}=\left\{\frac{1}{(n-1)!}\int_0^t e^{\alpha(t-\tau)}\,\tau^{n-1}\,e^{\alpha\tau}\,d\tau\right\}=\left\{\frac{t^n}{n!}e^{\alpha t}\right\}$$

(9) follows for $n+1$ too. □

Formulas (8) and (9) contain the *Heaviside algorithm* for solving initial value problems for linear differential equations with constant coefficients. Let $P(\lambda)=\alpha_0\lambda^n+\alpha_1\lambda^{n-1}+\cdots+\alpha_{n-1}\lambda+\alpha_n$ be a polynomial of n-th degree, i. e. with $\alpha_0\neq 0$. If $f\in\mathscr{C}(\Delta)$, then the initial value problem

$$P(d/dt)x=f,\quad x^{(k)}(0)=\gamma_k,\quad k=0,1,\ldots,n-1, \tag{10}$$

with given constants γ_k, can be reduced to the following equation:

$$P(s)x=f+Q(s) \tag{10'}$$

in $\mathsf{M}(\Delta)$, where $Q(\lambda)$ is the polynomial

$$Q(\lambda)=\sum_{j=1}^{n}\left(\sum_{k=j}^{n}\alpha_{n-j}\,\gamma_{k-j}\right)s^{j-1}$$

and its degree does not exceed $n-1$.

Since, according to Lemma 1, the element $P(s)$ of $\mathsf{M}(\Delta)$ is not a divisor of zero in $\mathsf{M}(\Delta)$, then the solution of (10′) in $\mathsf{M}(\Delta)$ is given by

$$x=\frac{1}{P(s)}f+\frac{Q(s)}{P(s)}\cdot \tag{11}$$

The fractions $1/P(s)$ and $Q(s)/P(s)$ can be represented as sums of partial fractions of the form (9), i. e. as sums of functions of $\mathscr{C}^\infty(\Delta)$. It is easily seen that (9) are functions of $\mathscr{C}^\infty(\Delta)$. It can easily be shown that x, given by (11), can be interpreted as function from $\mathscr{C}^n(\Delta)$. Hence (11) is the solution of the initial value problem (10). We will not go into further details, since we use this problem only as an illustration for a general scheme.

It is worth mentioning the following. The function $1/P(s)=\{G(t)\}$, which in our case is a quasi-polynomial, plays the role of the Green function of the problem (10), since for $\gamma_k=0$, $k=0, 1, 2, \ldots, n-1$, the solution of (10) can be represented in the form

$$x(t)=\int_0^t G(t-\tau)f(\tau)d\tau.$$

If by $g=\{g(t)\}$ we denote the solution of the initial value problem

$$P(d/dt)g(t)=1, \quad g^{(k)}(0)=0, \quad k=0, 1, \ldots, n-1, \tag{12}$$

then each solution of the problem (10) with zero initial conditions can be represented in the form

$$x(t)=\frac{d}{dt}\int_0^t g(t-\tau)f(\tau)\,d\tau, \tag{13}$$

reminding the representation (6) from sect. 1.1 of the multipliers of the Duhamel convolution algebra. Representation (13) is to be named a Duhamel representation. As we shall see later, it can be generalized for boundary value problems of the form

$$P(d/dt)x=f, \quad \Phi(x^{(k)})=0, \quad k=0, 1, \ldots, n-1,$$

with an arbitrary linear functional Φ in $\mathscr{C}(\Delta)$.

1.2.2. Interpretation as multiplier quotients. From Máté [75] and Struble [92] descends an interpretation of the Mikusiński ring as a quotient ring of the multiplier ring of the Duhamel convolution with respect to the multiplicative set of this ring, consisting of all non-trivial non-divisors of zero of this ring with respect to the operator multiplication.

Let us denote by $\mathfrak{M}=\mathfrak{M}_\Delta$ the multiplier ring of the Duhamel convolution algebra $\mathscr{C}(\Delta)$. We had studied this ring in sect. 1.1. Let us remind that an operator $M\colon \mathscr{C}(\Delta)\to\mathscr{C}(\Delta)$ is an element of $\mathfrak{M}$ iff the relation $M(f*g)=(Mf)*g$ holds for all $f, g\in\mathscr{C}(\Delta)$.

Let $\tilde{*}$ be another annihilators-free convolution of the Volterra integration operator l in $\mathscr{C}(\Delta)$, i. e. $\tilde{*}$ is a bilinear, commutative and associative operation in $\mathscr{C}(\Delta)$, for which the relation

$$l(f\tilde{*}g)=(lf)\tilde{*}g \tag{14}$$

holds for all $f, g\in\mathscr{C}(\Delta)$.

Theorem 1.2.1. *The multiplier ring of an arbitrary separately continuous and annihilators-free convolution $\tilde{*}$ of the Volterra integration operator l in $\mathscr{C}(\Delta)$ coincides with the multiplier ring $\mathfrak{M}=\mathfrak{M}_\Delta$ of the Duhamel convolution algebra $\mathscr{C}(\Delta)$.*

Proof. Let $\tilde{*}$ be a separately continuous convolution of l in $\mathscr{C}(\Delta)$ which is annihilators-free. If $M\colon \mathscr{C}(\Delta)\to\mathscr{C}(\Delta)$ is a multiplier of $\tilde{*}$, as in the proof of Lemma 7 from sect. 1.1, it can be proved that M commutes with l, i. e. that $Ml=lM$. According to Theorem 1.1.2, the operator M shall be a multiplier of the Duhamel convolution $*$ too, since as in the proof of Lemma 6 of sect. 1.1 it can easily be seen that M is a continuous linear operator.

Conversely, if $M \in \mathfrak{M}$, with analogous argument, but in reverse order, it can be seen that M is a multiplier of the convolution $\tilde{*}$. Let us now denote by $\mathfrak{N}=\mathfrak{N}_\Delta$ the multiplicative subset of $\mathfrak{M}$, consisting of all non-trivial non-divisors of zero of $\mathfrak{M}$ with respect to the operator multiplication. It is clear that $\mathfrak{N}$ is non-empty, because at least the identity I of $\mathscr{C}(\Delta)$ belongs to $\mathfrak{N}$. □

Lemma 3. *The sets of the non-divisors of zero of all separately continuous annihilators-free convolutions of the Volterra integration operator in $\mathscr{C}(\Delta)$ coincide with the set $\mathscr{C}_*(\Delta)$ of the non-divisors of the Duhamel convolution in $\mathscr{C}(\Delta)$.*

Proof. Indeed, if $\tilde{*}$ is an arbitrary separately continuous convolution of l in $\mathscr{C}(\Delta)$, let us denote by $\mathfrak{N}$ the set of the non-zero non-divisors of zero of $\tilde{*}$ in $\mathscr{C}_{\tilde{*}}(\Delta)$.

According to Theorem 1.1.4, we have $f \tilde{*} g=(d^2/dt^2)[\varrho * (f * g)]$. Therefore, if $f * g=0$, then $f \tilde{*} g=0$. Conversely, if $f \tilde{*} g=0$, then $\varrho * (f * g)=0$. It is easy to see that the requirement that $\tilde{*}$ is annihilators-free convolution is equivalent with the requirement ϱ to be a non-zero non-divisor of zero, in the Duhamel convolution algebra $\mathscr{C}(\Delta)$. Then $f * g=0$ and hence $\mathscr{C}_{\tilde{*}}(\Delta)=\mathscr{C}_*(\Delta)$, q. e. d. □

Definition. The quotient ring $\mathfrak{N}^{-1}\mathfrak{M}=(\mathfrak{M}\times\mathfrak{N})/_\sim$ with respect to the equivalence relation

$$(M, N)\sim(P, Q)\Leftrightarrow MQ=NP \tag{15}$$

in $\mathfrak{M}\times\mathfrak{N}$ is said to be the *ring of the multiplier quotients* of $\mathfrak{M}$.

It is obvious that the multiplier ring $\mathfrak{M}$ can be embedded in $\mathfrak{N}^{-1}\mathfrak{M}$ by the map

$$M \longrightarrow M/I. \tag{16}$$

The elements of $\mathfrak{N}^{-1}\mathfrak{M}$ will be denoted again as usual fractions.

Theorem 1.2.2. *The convolution quotients rings of all separately continuous and annihilators-free convolutions of the Volterra integration operator in $\mathscr{C}(\Delta)$ are isomorphic to the multiplier quotients ring $\mathfrak{N}^{-1}\mathfrak{M}$ of $\mathfrak{M}=\mathfrak{M}_\Delta$.*

Proof. Let $\tilde{*}$ be an arbitrary annihilators-free and separately continuous convolution of l in $\mathscr{C}(\Delta)$. If $f \in \mathscr{C}(\Delta)$, then by $(f\tilde{*})$ we shall denote the convolutional operator $(f\tilde{*})g=f\tilde{*}g$. According to Lemma 8 of sect. 1.1, the operator $(f\tilde{*})$ is a multiplier of the Duhamel convolution. If $(f/g)_{\tilde{*}}$ is an arbitrary element of the ring of the convolution quotients of $\mathscr{C}(\Delta)$ with respect to $\tilde{*}$, then the map

$$\left(\frac{f}{g}\right)_{\tilde{*}} \longrightarrow \frac{(f\tilde{*})}{(g\tilde{*})} \tag{17}$$

furnishes the desired isomorphism. In order to see this, one should make sure that (17) is well defined, i. e. that this correspondence is independent of the specific choice of f and g, and it is completely determined by the fraction $(f/g)_{\tilde{*}}$ alone. This can be settled easily. The correspondence (17) is one-to-one and each element of $\mathfrak{N}^{-1}\mathfrak{M}$ belongs to the image. Indeed, as we mentioned in the proof of Lemma 3, the element $\varrho=\{1\}\tilde{*}\{1\}$ is a non-trivial non-divisor of zero of $\tilde{*}$. Then each element M/N of $\mathfrak{N}^{-1}\mathfrak{M}$ is the image of the convolution quotient $(M\varrho/N\varrho)_{\tilde{*}}$ under the map (17). Then it is easy to see that (17) is an isomorphism between rings. □

Corollary. The convolution quotients rings of all annihilators-free and separately continuous convolutions of the Volterra integration operator l in $\mathscr{C}(\Delta)$ are isomorphic to the *Mikusiński ring.*

A similar but weaker statement is proved in [29].

Due to this isomorphism, we may use one and the same notation $\mathsf{M}(\Delta)$ for all these rings.

Remark. Although the Mikusiński ring $\mathsf{M}(\Delta)$ and the ring of the multiplier quotients are isomorphic, there exist subalgebras of the Duhamel convolution algebra $\mathscr{C}(\Delta)$ for which the analogous statement does not hold. Let $\Delta=[0, T]$ be a finite segment. Let us denote the subspace of $\mathscr{C}(\Delta)$ consisting of those functions which vanish in subsegments $[0, a]$ of $[0, T]$ by $\mathscr{C}_0(\Delta)$. Evidently, $\mathscr{C}_0(\Delta)$ is closed under the Duhamel convolution $*$. Each element $f\in\mathscr{C}_0(\Delta)$, however, is a divisor of zero. If $f\in\mathscr{C}(\Delta)$, then the convolution operator $(f*)x=f*x$ is a multiplier of the Duhamel convolution. Hence the ring of the multiplier quotients for the Duhamel convolution algebra $\mathscr{C}_0(\Delta)$ coincides with the Mikusiński ring. There is no convolution quotients ring of this Duhamel convolution algebra. This example is due to Máté [75].

1.2.3. The Mikusinski field. So far the special case $\Delta=[0, \infty)$ had been exclusively studied (see Mikusiński [79]). In this case $\mathscr{C}(\Delta)$ is a field — the Mikusiński field. Here we discuss some specific features of this case.

Remarkable elements of the *Mikusiński field* are the so-called "translation" operators.

Let $f(t)$ be an arbitrary function of $\mathscr{C}([0, \infty))$ with $f(0)=0$. For arbitrary $\lambda\geqq 0$ we define f_λ by

$$f_\lambda(t)=\begin{cases} 0, & 0\leqq t\leqq\lambda, \\ f(t-\lambda), & \lambda\leqq t. \end{cases} \tag{18}$$

Lemma 4. *The convolution quotient f_λ/f does not depend on the function $f\in\mathscr{C}([0, \infty))$ with $f(0)=0$.*

Proof. Let $g(t)$ be another function of $\mathscr{C}([0, \infty))$ with $g(0)=0$. If $t\leqq\lambda$, then we have $f_\lambda*g(t)=0$ and $f*g_\lambda(t)=0$. Consider the case $t>\lambda$. Then

$$(f_\lambda*g)(t)=\int_0^{t-\lambda} f(t-\lambda-\sigma)\,g(\sigma)d\sigma=\int_0^{t-\lambda} f(\sigma)g(t-\lambda-\sigma)d\sigma$$

and hence $f_\lambda*g=f*g_\lambda$. □

Definition. The convolution quotient e_λ, defined for $\lambda\geqq 0$ by

$$e_\lambda=\frac{f_\lambda}{f} \quad \text{with } f\in\mathscr{C}([0, \infty)),\ f(0)=0, \tag{19}$$

and for $\lambda<0$ by $e_\lambda=1/e_{-\lambda}$, is said to be a *translation operator.*

It is easy to prove that $e_\lambda e_\mu=e_{\lambda+\mu}$ for all λ, μ.

It is quite natural to attempt to develop an analysis for functions of a real variable with the values in the Mikusiński field. But difficulties arise even for the definitions of basic notions as the notion of derivative.

If the real variable λ changes on a segment $[a, b]$, then the derivative of a function $f: [a, b] \to \mathsf{M}$ of the form $f(\lambda)=\{\varphi(t, \lambda)\}/\{\psi(t, \lambda)\}$, where $\varphi(t, \lambda)$ and $\psi(t, \lambda)$ are defined on $[0, \infty)\times[a, b]$ and have continuous partial derivatives, can be defined as

$$f'(\lambda)=\frac{\{\varphi_\lambda(t, \lambda)\} * \{\psi(t, \lambda)\}-\{\varphi(t, \lambda)\} * \{\psi_\lambda(t, \lambda)\}}{\{\psi(t, \lambda)\}^{*2}}. \tag{20}$$

It is an open problem to prove or disprove "the fundamental theorem of the integral calculus": If $f'(\lambda)=0$ for $a\leq\lambda\leq b$, then $f(\lambda)=\text{const}$. It can be stated in classical terms in the following way: Does the identity

$$\int_0^t \varphi_\lambda(t-\tau, \lambda)\psi(\tau, \lambda)d\tau=\int_0^t \varphi(t-\tau, \lambda)\psi_\lambda(\tau, \lambda)d\tau,$$

where $\varphi(t, \lambda)$ and $\psi(t, \lambda)$ satisfy the same conditions as above, imply the identity

$$\int_0^t \varphi(t-\tau, \lambda)\psi(\tau, a)d\tau=\int_0^t \psi(t-\tau, \lambda)\varphi(\tau, a)d\tau$$

in $[0, \infty)\times[a, b]$? It may be additionally assumed that $\psi(t, \lambda)$ does not vanish identically on t for any $\lambda\in[a, b]$. *(The Mikusiński's problem).*

Later on (in Chapter 2) we give an example of a convolution for which the analogous problem is solved positively without any difficulties.

1.3. CONVOLUTIONS OF LINEAR ENDOMORPHISMS

The relation between the Duhamel convolution and the Volterra integration operator suggests a general notion of convolution. If $\mathfrak{X}$ is a linear space and if $L: \mathfrak{X}\to\mathfrak{X}$ is an endomorphism of $\mathfrak{X}$, then under a convolution of L it is natural to understand every bilinear, commutative and associative operation $*: \mathfrak{X}\times\mathfrak{X}\to\mathfrak{X}$ such that the relation $L(x*y)=(Lx)*y$ is fulfilled for all $x, y\in\mathfrak{X}$. This definition proposed by the author [30] in 1966 is only an inverting of the definition of a multiplier of commutative and associative algebra. In a sense the purpose of convolutional calculus is converse to the purpose of the multiplier theory. The basic problem of the latter theory is to characterize the multipliers of a given convolutional algebra. The basic problem of a convolutional calculus is to characterize the convolutions of a given endomorphism of a linear space. The problem to find at least one such non-trivial convolution is non-trivial. It is solved only in too special cases. Each of these cases is interesting on its own right, due to the fact that by means of the corresponding convolutions one can find non-trivial analytical results using only simple algebraic considerations.

Leaving the study of such new convolutions for the following two chapters, here we shall study the general properties of the convolutions of a linear space endomorphism and some algebraic constructions based on them.

1.3.1. Definition of a convolution of linear endomorphism. Let $\mathfrak{X}$ be a linear space. In general considerations we may assume that $\mathfrak{X}$ is a linear space over an arbitrary field with characteristic 0, but in fact in the following we shall restrict ourselves with linear spaces over the field C of the complex numbers.

Let $L: \mathfrak{X} \to \mathfrak{X}$ be a fixed endomorphism of $\mathfrak{X}$, i. e. a linear operator mapping $\mathfrak{X}$ into itself. In sect. 1.1 we have considered the space $\mathscr{C}(\Delta)$ of the continuous functions on an interval Δ containing the zero point, and as an endomorphism in it we have taken the Volterra integration operator l. The Duhamel convolution

$$f * g = \left\{ \int_0^t f(t-\tau) g(\tau) d\tau \right\}$$

converts $\mathscr{C}(\Delta)$ into commutative and associative algebra. The operator l is the convolutional operator $\{1\}*$, i. e. $lf = \{1\} * f$.

This example suggests a definition of convolution of an endomorphism $L: \mathfrak{X} \to \mathfrak{X}$. Under a convolution of L we can understand every bilinear, commutative and associative operation $*: \mathfrak{X} \times \mathfrak{X} \to \mathfrak{X}$ in $\mathfrak{X}$ for which L can be represented as a convolutional operator, i. e. as $L = r*$ with some $r \in \mathfrak{X}$. But the relation $Lx = r * x$ is too restrictive. In [30] the author replaced the condition L to be representable in the operation $x * y$ by an element of the space by a weaker restriction L to be a multiplier of the algebra under consideration. Let us state now the corresponding definition in full.

Definition 1. Let $\mathfrak{X}$ be a linear space and let $L: \mathfrak{X} \to \mathfrak{X}$ be a linear operator. A bilinear, commutative and associative operation $*: \mathfrak{X} \times \mathfrak{X} \to \mathfrak{X}$ is said to be a *convolution of the linear operator L* iff the relation

(1) $$L(x * y) = (Lx) * y$$

is fulfilled for all $x, y \in \mathfrak{X}$.

The bilinearity can be expressed in the following way: For all $x_1, x_2, y_1, y_2 \in \mathfrak{X}$ and for arbitrary constants $\alpha_1, \alpha_2, \beta_1, \beta_2$ the bilinearity relation

(2) $$(\alpha_1 x_1 + \alpha_2 x_2) * (\beta_1 y_1 + \beta_2 y_2) = \alpha_1 \beta_1 (x_1 * y_1) + \alpha_1 \beta_2 (x_1 * y_2) + \alpha_2 \beta_1 (x_2 * y_1) + \alpha_2 \beta_2 (x_2 * y_2)$$

is fulfilled.

The commutativity and associativity can be expressed by the identities

(3) $$x * y = y * x$$

and

(4) $$(x * y) * z = x * (y * z).$$

Of course, we shall be interested in non-trivial convolutions only, i. e $x * y = 0$ not for all $x, y \in \mathfrak{X}$. It is necessary to make a further restriction on the possible convolutions of the operator $L: \mathfrak{X} \to \mathfrak{X}$. We should exclude the convolutions with annihilators. An element $a \in \mathfrak{X}$, $a \neq 0$, is said to be an annihilator of a convolution $x * y$ in $\mathfrak{X}$ iff $a * x = 0$ for each $x \in \mathfrak{X}$.

Definition 2. A convolutional algebra $(\mathfrak{X}, *)$ is said to be *annihilators-free* if $x * y = 0$ for each $x \in \mathfrak{X}$ implies $y = 0$.

It is not difficult to show a non-trivial convolution of the Volterra integration operator L having annihilators. Let $f*g$ be the Duhamel convolution in $\mathscr{C}([0, 1])$. If r is a divisor of zero of the Duhamel convolution, then the operation

$$f \tilde{*} g = r*(f*g)$$

is a convolution of L in $\mathscr{C}([0, 1])$, which is not annihilators-free. Indeed, if $q \in \mathscr{C}([0, 1])$, $q \neq 0$, is such that $r*q=0$, then $f \tilde{*} q=0$ for all $f \in \mathscr{C}([0, 1])$.

In all further considerations we shall assume that the convolutions considered are annihilators-free, but we should not exclude at all the non-trivial convolutions with annihilators. It may happen the restriction of such a convolution to a subspace of the basic space $\mathfrak{X}$ to be an annihilators-free convolution.

If we have at our disposal a given convolutional algebra, then the first problem connected with it is to find all the endomorphisms $M: \mathfrak{X} \to \mathfrak{X}$ of the linear space of the algebra $\mathfrak{A}=(\mathfrak{X}, *)$, such that for them the multiplication $*$ in $\mathfrak{X}$ to be a convolution of M in the sense of Definition 1.

Two classes of such operators can easily be shown. If $m \in \mathfrak{X}$ is arbitrarily chosen, then the convolutional operator $M=m\,*$, defined by

$$Mx=m*x, \tag{5}$$

has the desired property. It $\lambda \in \mathbf{C}$ is an arbitrary constant, then the operator $\Lambda=[\lambda]$, defined by

$$\Lambda x=[\lambda]x=\lambda x \tag{6}$$

is a linear operator in $\mathfrak{X}$ with a convolution $*$ in $\mathfrak{X}$.

Definition 3. If $\mathfrak{A}=(\mathfrak{X}, *)$ is a *convolutional algebra,* then a linear operator $M: \mathfrak{X} \to \mathfrak{X}$ is said to be a *multiplier of* $\mathfrak{A}$, if the relation

$$M(x*y)=(Mx)*y \tag{7}$$

is fulfilled for all $x, y \in \mathfrak{X}$.

Now we shall establish some properties of the multipliers of annihilators-free algebras.

Lemma 1. *Let $\mathfrak{A}=(\mathfrak{X}, *)$ be a commutative and associative annihilators-free algebra and let $M: \mathfrak{X} \to \mathfrak{X}$. Then condition* (7) *is equivalent to*

$$(Mx)*y=x*(My). \tag{8}$$

Proof. It is obvious that (7) implies (8), provided we use the commutativity of $*$. Let us show the converse implication. From (7) we have $[M(x*y)]*z=(x*y)*(Mz)$. But $[(Mx)*y]*z=[x*(My)]*z=x*[(My)*z]=x*[y*(Mz)]$, and hence $[M(x*y)-(Mx)*y]*z=0$. From the assumption that $\mathfrak{A}$ is annihilators-free it follows (7). □

Lemma 2. *For an annihilators-free commutative and associative algebra $\mathfrak{A}$ with multiplication $*$ the relation* (1) *is equivalent to relation*

$$L(x*x)=(Lx)*x \tag{9}$$

for all $x \in \mathfrak{A}$.

Proof. It is obvious that (1) implies (9). Now, let us prove the converse. If $x, y \in \mathfrak{A}$ are arbitrary, then from (9) it follows $L[(x+y)^{*2}]=(x+y) * L(x+y)$. After some transformations, this equality takes the form

$$(10) \qquad 2L(x*y)=(Lx)*y+x*(Ly).$$

By means of (10) we transform the expression $4L(x*y*z)$ in two ways, using the representations $L(x*y*z)=L[x*(y*z)]=L[(x*y)*z]$. Thus we get $[(Lx)*y]*z=[x*(Ly)]*z$. But z can be an arbitrary element of $\mathfrak{A}$ and hence $(Lx)*y=x*(Ly)$ since the algebra $\mathfrak{A}$ is annihilators-free. □.

Theorem 1.3.1 (Máté [75]). *The multipliers of an annihilators-free algebra $\mathfrak{A}=(\mathfrak{X}, *)$ are linear operators in the linear space $\mathfrak{X}$ of the algebra. They form a commutative algebra with respect to the operator multiplication.*

Proof. The first part of the statement can be proved in the same way as in the proof of the corresponding proposition of the Duhamel convolution algebra (Lemma 6 of 1.1.2). However, for completeness we shall repeat the argument. If $x, y, z \in \mathfrak{X}$ are arbitrary, and if $\alpha, \beta \in \mathbf{C}$ are arbitrary constants, then for each multiplier $M: \mathfrak{X} \to \mathfrak{X}$ of the algebra $\mathfrak{A}=(\mathfrak{X}, *)$ we have

$$[M(\alpha x+\beta y)-(\alpha Mx+\beta My)]*z=[M(\alpha x+\beta y)]*z$$
$$-(\alpha Mx)*z-(\beta My)*z=(\alpha x+\beta y)*Mz-\alpha(Mx*z)-\beta(My*z)$$
$$=(\alpha x)*Mz+(\beta y)*Mz-\alpha(x*Mz)-\beta(y*Mz)=0.$$

Since $\mathfrak{A}$ is annihilators-free, then $M(\alpha x+\beta y)=\alpha Mx+\beta My$, i. e. M is a linear operator in $\mathfrak{X}$.

Let $M: \mathfrak{X} \to \mathfrak{X}$ and $N: \mathfrak{X} \to \mathfrak{X}$ be two arbitrary multipliers of $\mathfrak{A}$, and let x, y be arbitrary elements of $\mathfrak{X}$. Then $(MNx-NMx)*y=(MNx)*y-(NMx)*y =M[(Nx)*y]-[(Mx)*(Ny)]=M[x*(Ny)]-M[x*(Ny)]=0$ and hence $MNx =NMx$, i. e. every two multipliers of $\mathfrak{A}$ commute.

From relation (7) it is obvious that the product MN of two multipliers M and N of $\mathfrak{A}$ is a multiplier of $\mathfrak{A}$ too. It is not less evident that if $\alpha \in \mathbf{C}$, then αM is a multiplier of $\mathfrak{A}$, provided M is a multiplier. □

Theorem 1.3.2. *If $\mathfrak{A}$ is an annihilators-free convolutional algebra, and $\mathfrak{M}$ is the algebra of its multipliers, then the mappings $\mathfrak{A} \to \mathfrak{M}$ and $\mathbf{C} \to \mathfrak{M}$, defined as $m \mapsto m*$ and $\lambda \mapsto [\lambda]$ by the correspondences $(m*)f =m*f$ and $[\lambda]f=\lambda f$, according to (5) and (6), are isomorphic embeddings of algebras.*

Proof. Since $\mathfrak{A}$ is annihilators-free, then the map $m \to m*$ is an injection. From the associativity of $*$ it follows $(mn)*=(m*)(n*)$. The identities $(m+n)*=(m*)+(n*)$ and $(\lambda n)*=\lambda(m*)$ for an arbitrary $\lambda \in \mathbf{C}$ are evident. The second part of the statement is not less evident. If $\Lambda=[\lambda]$ and $M=[\mu]$ with arbitrary $\lambda, \mu \in \mathbf{C}$, then $\Lambda M=[\lambda\mu]$ and $\Lambda+M=[\lambda+\mu]$. Therefore the maps $\mathfrak{A} \to \mathfrak{M}$ and $\mathbf{C} \to \mathfrak{M}$ defined by (5) and (6) are isomorphic embeddings of algebras. □

The embedding $\mathbf{C} \to \mathfrak{M}$ defined by $\lambda \mapsto [\lambda]$ allows us to reduce the three operations in the multiplier algebra $\mathfrak{M}$ to two: the addition $M+N$ and the multiplication MN. The third operation, multiplication αM of a multiplier M by a constant α, can be reduced to the operator multiplication due to the identity

$$(11) \qquad \alpha M=[\alpha]M.$$

If $\alpha \in \mathbf{C}$ and $x \in \mathfrak{A}$ are arbitrary, then the map

(12) $$\alpha x \to [\alpha](x*)$$

shows that to the product of a number with an element of $\mathfrak{A}$ there corresponds the product of the respective multipliers of $\mathfrak{M}$. This shows that we can look on $\mathfrak{A}$ and $\mathbf{C}$ as on parts of one of the same algebraic system, the ring $\mathfrak{M}$ of the multipliers of $\mathfrak{A}$. Thus no misunderstanding would arise if we drop the star in $x*$, writing simply x for this convolutional operator, or the brackets in $[\alpha]$, writing simply α for this multiplier operator. Then expressions like $\alpha+x$ with $\alpha \in \mathbf{C}$ and $x \in \mathfrak{X}$, should be understood as

(13) $$\alpha+x \overset{\text{def}}{=} [\alpha]+(x*).$$

But one should have in mind that this is justified only when the algebra $\mathfrak{A}$ is annihilators-free.

The multiplier ring $\mathfrak{M}$ of an annihilators-free algebra $\mathfrak{A}$ is evidently a ring with unit. We shall denote the unit of $\mathfrak{M}$ by 1 since it is the multiplier 1, i. e. the identity of $\mathfrak{M}$. Looking on $\mathfrak{A}$ as on a part of $\mathfrak{M}$, we can write fg instead of $f*g$ which we shall do often further.

1.3.2. Divisors of zero of a convolution of a linear space endomorphism. Let L be an endomorphism of a linear space $\mathfrak{X}$, and let $*$ be a convolution of L in $\mathfrak{X}$. We are interested mainly in annihilators-free convolutions, i. e. in multiplications $*$ in $\mathfrak{X}$ having the property the fulfilling of $x*y=0$ for each $x \in \mathfrak{X}$ to imply $y=0$. Thus we by no means exclude the convolutions with divisors of zero. An element $a \in \mathfrak{X}$ is said to be a divisor of zero of a convolution $*$ in $\mathfrak{X}$ iff $a \neq 0$ and there exists an element $b \neq 0$ such that $a*b=0$.

In Lemmas 2 and 3 in sect. 1.1 we have characterized the divisors of zero of the Duhamel convolution

$$(f*g)(t)=\int_0^t f(t-\tau)g(\tau)d\tau$$

in $\mathscr{C}(\Delta)$, where Δ is an interval containing the point $t=0$. According to the well-known Titchmarsh theorem, the Duhamel convolution algebra $(\mathscr{C}(\Delta), *)$ has no divisors of zero when $\Delta=[0,\infty)$.

Now we shall prove a useful theorem which gives some information about divisors of zero of an arbitrary convolution of linear operator L: $\mathfrak{X}\to\mathfrak{X}$.

Theorem 1.3.3. *Let L: $\mathfrak{X}\to\mathfrak{X}$ be a linear operator. If for a positive integer n and for a complex number λ the subspace*

(14) $$\mathfrak{X}_{\lambda,n}=\ker\,(L-\lambda)^n=\{x:\ x\in\mathfrak{X},\ (L-\lambda)^n x=0\}$$

is neither $\{0\}$ nor the whole space $\mathfrak{X}$, then each non-zero element of $\mathfrak{X}_{\lambda,n}$ is a divisor of zero of each convolution of L in $\mathfrak{X}$.

Proof. Let $a \in \mathfrak{X}_{\lambda,n}$ and $a \neq 0$. According to the hypothesis there is a $b \in \mathfrak{X}$, such that $b \notin \mathfrak{X}_{\lambda,n}$. Then $b \neq 0$. If $*$ is an arbitrary convolution of L in $\mathfrak{X}$, then let us multiply the equality $(L-\lambda)^n a=0$ by b. Thus we get

$$[(L-\lambda)^n a]*b=[(L-\lambda)^n b]*a=0.$$

But $(L-\lambda)^n b \neq 0$ since $b \notin \mathfrak{X}_{\lambda,n}$. Hence a is a divisor of zero of the convolution considered. □

Corollary. Every eigenelement of L is a divisor of zero of each of its convolutions.

This is the theorem for $n=1$.

Now we shall show that each two subspaces $\mathfrak{X}_{\lambda,n}$ and $\mathfrak{X}_{\mu,m}$ of the form (14) are "convolutionally orthogonal", i. e.

$$\mathfrak{X}_{\lambda,n} * \mathfrak{X}_{\mu,m} = \{0\} \text{ for all } \lambda \neq \mu.$$

Theorem 1.3.4. *Let $\lambda \neq \mu$ and let $x*y$ be an arbitrary convolution of L in $\mathfrak{X}$. Then the convolutional product of each $u \in \mathfrak{X}_{\lambda,n} = \ker(L-\lambda)^n$ with each $v \in \mathfrak{X}_{\mu,m} = \ker(L-\mu)^m$ is equal to* 0.

Proof. We shall prove the statement by induction. Let $m=n=1$. If one of the subspaces $\mathfrak{X}_{\lambda,1} = \ker(L-\lambda)$ and $\mathfrak{X}_{\mu,1} = \ker(L-\mu)$ is the zero space, then the assertion is true. Now let us assume that they are both non-trivial. Let $u \neq 0$ with $(L-\lambda)u=0$ and $(L-\mu)v=0$ with $v \neq 0$. If we multiply $Lu=\lambda u$ by v and use the convolutional relation $L(u*v)=(Lu)*v$, we get $L(u*v)=\lambda(u*v)$. In the same way, multiplying $Lv=\mu v$ by u, we get $L(u*v) = \mu(u*v)$, and $u*v=0$ since $\lambda \neq \mu$.

Let $(L-\lambda)^n u=0$ and k be the smallest integer for which $(L-\lambda)^k u=0$. Then $(L-\lambda)^{k-1} u \neq 0$. Let l be the smallest integer with $(L-\mu)^l v=0$. Then $(L-\mu)^{l-1} v \neq 0$. Let us consider the finite sequences

(15) $$u_1=(L-\lambda)^{k-1}u,\ u_2=(L-\lambda)^{k-2}u, \ldots,\ u_k=u$$

and

(15') $$v_1=(L-\mu)^{l-1}v,\ v_2=(L-\mu)^{l-2}v, \ldots,\ v_l=v.$$

Since $(L-\lambda)u_1=0$ and $(L-\mu)v_1=0$, then as we have proved, $u_1*v_1=0$. The proof that $u_i*v_j=0$ for each i with $1 \leq i \leq k$ and for each j with $1 \leq j \leq l$ proceeds by induction on $i+j$. Let us assume that the assertion is already proved for all i and j with sum less than or equal to an integer $p<k+l$, i. e. with $i+j \leq p$. Let now $i+j=p+1$. We should prove that $w \overset{\text{def}}{=} u_i*v_j=0$. Indeed we have

$$(L-\lambda)w=[(L-\lambda)u_i]*v_j=u_{i-1}*v_j=0$$

when $i>1$. If $i=1$, then $(L-\lambda)u_i=0$ and $(L-\lambda)w=0$. In the same way it is seen that $(L-\mu)w=0$. But we have assumed that $\lambda \neq \mu$. Hence $w=u_i*v_j=0$ q. e. d. The above proof resembles a proof of a theorem of linear algebra (see Faddeev, Faddeeva [57], p. 102). □

Let λ be an arbitrary eigenvalue of a linear operator $L: \mathfrak{X} \to \mathfrak{X}$. The subspaces $\mathfrak{X}_{\lambda,n}$, defined by (14), form an increasing sequence of subspaces of $\mathfrak{X}$, i. e.

(16) $$\mathfrak{X}_{\lambda,1} \subset \mathfrak{X}_{\lambda,2} \subset \ldots \subset \mathfrak{X}_{\lambda,k} \subset \mathfrak{X}_{\lambda,k+1}.$$

If it happens two successive members of this sequence to coincide, then from this place on the sequence stabilizes. Let k be the first number with $\mathfrak{X}_{\lambda,k}=\mathfrak{X}_{\lambda,k+1}$. We have $\mathfrak{X}_{\lambda,k}=\mathfrak{X}_{\lambda,k+1}=\mathfrak{X}_{\lambda,k+2}=\ldots$ This integer k is said to be the *spectral multiplicity* of the eigenvalue λ (see Edwards [55], sect. 9.8). If (16) is a strictly increasing sequence, then we define the spectral multiplicity of λ as ∞.

Obviously, each of the subspaces $\mathfrak{X}_{\lambda,n}$ is an ideal in the convolutional algebra $(\mathfrak{X}, *)$ for every convolution $*$ of L in $\mathfrak{X}$. The subspace $\mathfrak{X}_{\lambda,k}$ has the simplest structure, if the eigenvalue λ is simple, i. e. such that the eigenspace $\mathfrak{X}_{\lambda,1}=\ker(L-\lambda)$ is one-dimensional.

Theorem 1.3.5. *Let λ be a simple eigenvalue of the linear operator $L: \mathfrak{X}\to\mathfrak{X}$, having a finite spectral multiplicity k. If $*$ is a convolution of L in $\mathfrak{X}$, which is annihilators-free in $\mathfrak{X}_{\lambda,k}$, then there exists an element $u\in\mathfrak{X}_{\lambda,k}$, such that $x*u=x$ for each $x\in\mathfrak{X}_{\lambda,k}$.*

Proof. Since $\mathfrak{X}_{\lambda,k}$ is an ideal in every convolutional algebra $(\mathfrak{X}, *)$ of $\mathfrak{X}$, then the restriction of a convolution of L in $\mathfrak{X}$ to $\mathfrak{X}_{\lambda,k}$ is an inner operation for $\mathfrak{X}_{\lambda,k}$. By the hypothesis, the convolution $*$ is annihilators-free in $\mathfrak{X}_{\lambda,k}$. Let us consider this restriction to $\mathfrak{X}_{\lambda,k}$ of the given convolution $x*y$.

Let $\tilde{u}\in\mathfrak{X}_{\lambda,k}$ be an arbitrary element with $(L-\lambda)^{k-1}\tilde{u}\neq 0$. Let us form the sequence corresponding to (15):

$$\tilde{u}_1=(L-\lambda)^{k-1}\tilde{u},\ \tilde{u}_2=(L-\lambda)^{k-2}\tilde{u},\ldots,\ \tilde{u}_k=\tilde{u}.$$

Now we shall find the multiplication table for the products $\tilde{u}_i*\tilde{u}_j$ $i,j=1,2,\ldots,k$. Since

$$\tilde{u}_i*\tilde{u}_j=(L-\lambda)^{2k-i-j}(\tilde{u}*\tilde{u}), \tag{17}$$

it is clear that the product depends on the sum of the subscripts i and j only, and

$$\tilde{u}_i*\tilde{u}_j=0,\quad i+j\leq k. \tag{17'}$$

Since the elements $\tilde{u}_1, \tilde{u}_2, \ldots, \tilde{u}_k$ form a basis of the k-dimensional space $\mathfrak{X}_{\lambda,k}$ (see Faddeev, Faddeeva [57], p. 77), then

$$\tilde{u}*\tilde{u}=C_1\tilde{u}_1+C_2\tilde{u}_2+\cdots+C_k\tilde{u}_k$$

with some constants $C_1, C_2, \ldots, C_k$. Then we obtain

$$\begin{aligned}\tilde{u}_k*\tilde{u}_1 &= C_k\tilde{u}_1,\\ \tilde{u}_k*\tilde{u}_2 &= C_{k-1}\tilde{u}_1+C_k\tilde{u}_2,\\ &\ldots\ldots\ldots\\ \tilde{u}_k*\tilde{u}_k &= C_1\tilde{u}_1+C_2\tilde{u}_2+\cdots+C_k\tilde{u}_k.\end{aligned} \tag{18}$$

From the assumption that the convolution $*$ is annihilators-free it follows that $C_k\neq 0$. Indeed if $C_k=0$, then we should have $\tilde{u}_k*\tilde{u}_1=\tilde{u}_{k-1}*\tilde{u}_1=\cdots=\tilde{u}_1*\tilde{u}_1=0$, i. e. $\tilde{u}_1$ should be an annihilator in $\mathfrak{X}_{\lambda,k}$. Hence $C_k\neq 0$. Now we are looking for an element $u\in\mathfrak{X}_{\lambda,n}$, such that $u*u=u$. For this we search coefficients $A_1, A_2, \ldots, A_k$ such that $u=A_1\tilde{u}_1+A_2\tilde{u}_2+\cdots+A_k\tilde{u}_k$. Using the formulas (17), (17′) and (18) we obtain the following triangle system for the coefficients from the requirement $u*\tilde{u}_k=\tilde{u}_k$:

$$\begin{aligned}&A_kC_k=1,\\ &A_{k-1}C_k+A_kC_{k-1}=0,\\ &\ldots\ldots\ldots\\ &A_1C_k+A_2C_{k-1}+\cdots+A_kC_1=0.\end{aligned}$$

These simultaneous equations always have a solution for $A_1, A_2, \ldots, A_k$. Thus the existence of an element $u \in \mathfrak{X}$ with $u * u = u$ is proved. Then $u * \tilde{u}_j = u_j$ for each j, with $1 \leq j \leq k$. Indeed from $u_j = (L-\lambda)^{k-j}\tilde{u}_k$ it follows $u * \tilde{u}_j = (L-\lambda)^{k-j}(u * \tilde{u}_k) = (L-\lambda)^{k-j}\tilde{u}_k = u_j$. Since $u_1, u_2, \ldots, u_k$ is a basis in $\mathfrak{X}_{\lambda,k}$, then $u * x = x$ for each $x \in \mathfrak{X}_{\lambda,k}$. Thus the theorem is proved.□

Corollary. For every simple eigenvalue λ of the operator $L: \mathfrak{X} \to \mathfrak{X}$ which has finite spectral multiplicity μ and for every convolution of L in $\mathfrak{X}$ which is annihilators-free in $\mathfrak{X}_{\lambda,k}$, there exists a *multiplier projector* P: $\mathfrak{X} \to \mathfrak{X}_{\lambda,k}$ of the form $Px = u * x$, where $u \in \mathfrak{X}_{\lambda,k}$ is an element with the property $u * u = u$.

The operator $Q = 1 - P$ is also a multiplier projector, and the space $\mathfrak{X}$ can be represented as the direct sum

$$\mathfrak{X} = P(\mathfrak{X}) \oplus (1-P)(\mathfrak{X}) = \mathfrak{X}_{\lambda,k} \oplus Q(\mathfrak{X}).$$

A slightly different approach is proposed by Bozhinov [15] and [16]

1.3.3. Convolutions of similar operators. The main way to find new convolutions from known ones is the similarity method. The use of integral transforms for developing operational calculi is in fact use of the similarity method. Though the basic idea lying under it is a very simple one, it is quite effective. Strange is only the fact that it had not been used in many cases where it is well applicable.

Definition 4. A linear operator $L: \mathfrak{X} \to \mathfrak{X}$ is said to be *similar* to a linear operator $\tilde{L}: \tilde{\mathfrak{X}} \to \tilde{\mathfrak{X}}$ if there exists a linear isomorphism $T: \mathfrak{X} \to \tilde{\mathfrak{X}}$ such that

$$TL = \tilde{L}T. \tag{19}$$

It is also said that T is a *similarity* of L to $\tilde{L}$. Relation (19) expresses the commutativity of the simple diagram

$$\begin{array}{ccc} \mathfrak{X} & \xrightarrow{L} & \mathfrak{X} \\ T\downarrow & & \downarrow T \\ \tilde{\mathfrak{X}} & \xrightarrow{L} & \tilde{\mathfrak{X}} \end{array}$$

and it can be written in the form

$$L = T^{-1}\tilde{L}T. \tag{19'}$$

Basic for the convolutional calculus, at least as it is understood in this book, is the following, almost evident proposition:

Theorem 1.3.6. *If $T: \mathfrak{X} \to \tilde{\mathfrak{X}}$ is an isomorphism of linear space $\mathfrak{X}$ onto a linear space $\tilde{\mathfrak{X}}$, and $\tilde{L}: \tilde{\mathfrak{X}} \to \tilde{\mathfrak{X}}$ is a linear operator in $\mathfrak{X}$ with a convolution $\tilde{*}: \tilde{\mathfrak{X}} \times \tilde{\mathfrak{X}} \to \tilde{\mathfrak{X}}$, then the operation*

$$x * y = T^{-1}(Tx \,\tilde{*}\, Ty) \tag{20}$$

is a convolution of the operator $L = T^{-1}\hat{L}T$ in $\mathfrak{X}$.

Proof. It is obvious that operation (20) is a bilinear and commutative one, since $\tilde{*}$ is such an operation. Let us prove the associativity of (20). We have

$$(x*y)*z=T^{-1}[TT^{-1}(Tx*Ty)\tilde{*}Tz]$$
$$=T^{-1}[(Tx\tilde{*}Ty)\tilde{*}Tz]=T^{-1}[Tx\tilde{*}(Ty\tilde{*}Tz)]$$
$$=T^{-1}[Tx\tilde{*}TT^{-1}(Ty\tilde{*}Tz)]=x*(y*z),$$

where only the associativity of $\tilde{*}$ is used.

It remains to establish the convolutional relation $L(x*y)=(Lx)*y$. We have

$$L(x*y)=T^{-1}\tilde{L}T.T^{-1}(Tx\tilde{*}Ty)$$
$$=T^{-1}\tilde{L}(Tx\tilde{*}Ty)=T^{-1}[\tilde{L}(Tx)\tilde{*}Ty]$$
$$=T^{-1}[TLx\tilde{*}Ty]=(Lx)*y,$$

where the convolutional property $\tilde{L}(\tilde{x}\tilde{*}\tilde{y})=(\tilde{L}\tilde{x})\tilde{*}\tilde{y}$ of $\tilde{L}$ is used. □

Under a more restrictive definition of convolution, a similar theorem is proved by Meller [76].

Corollary 1. Under the hypothesis of Theorem 1.3.6, the convolutional algebras $(\mathfrak{X}, *)$ and $(\tilde{\mathfrak{X}}, \tilde{*})$ are isomorphic.

Indeed, from definition (20)

$$T(x*y)=(Tx)\tilde{*}(Ty).$$

This shows that $T:\mathfrak{X}\to\tilde{\mathfrak{X}}$ is an isomorphism not only of linear spaces, but of algebras too.

Therefore, if one of the algebras $(\mathfrak{X}, *)$ and $(\tilde{\mathfrak{X}}, \tilde{*})$ is annihilators-free, then the other is annihilators-free too.

Corollary 2. Let $T:\mathfrak{X}\to\tilde{\mathfrak{X}}$ be a linear and invertible map of $\mathfrak{X}$ in $\tilde{\mathfrak{X}}$, and let $\tilde{L}:\tilde{\mathfrak{X}}\to\tilde{\mathfrak{X}}$ be a linear operator in $\tilde{\mathfrak{X}}$ with a convolution $\tilde{*}$: $\tilde{\mathfrak{X}}\times\tilde{\mathfrak{X}}\to\tilde{\mathfrak{X}}$. If the image-space $T(\mathfrak{X})$ of $\mathfrak{X}$ under T is closed with respect to the operation $\tilde{*}$, then operation (20) is a convolution of the operator $L=T^{-1}\tilde{L}T$ in $\mathfrak{X}$.

Under the closedness of $T(\mathfrak{X})$ with respect to $\tilde{*}$ it is understood that if $\tilde{x}, \tilde{y}\in T(\mathfrak{X})$, then $\tilde{x}\tilde{*}\tilde{y}\in T(\mathfrak{X})$. The assertion is an immediate consequence of Theorem 1.3.6, provided we substitute $\tilde{\mathfrak{X}}$ by $T(\mathfrak{X})$.

Example. Let $\mathfrak{X}=\mathscr{C}([a,b])$, and $\varphi(x)$ be a strictly increasing function in $[a,b]$ with $\varphi(a)=0$. We are looking for a non-trivial convolution of the linear operator

$$Lf(x)=\int_a^x f(\xi)d\varphi(\xi)$$

in $\mathscr{C}([a,b])$.

The transformation

$$T: f(x)\to f[\varphi^{-1}(t)]$$

is an isomorphism of the linear space $\mathscr{C}([a,b])$ onto $\mathscr{C}([0,\varphi(b)])$. It is a similarity of the operator L to the Volterra integration operator l, since

$$TLf(x)=\int_a^{\varphi^{-1}(t)} f(\xi)d\varphi(\xi)=\int_0^t f[\varphi^{-1}(\tau)]d\tau=lTf.$$

But the Duhamel convolution

$$(\tilde{f} \,\tilde{*}\, \tilde{g})(t)=\int_0^t \tilde{f}(t-\tau)\tilde{g}(\tau)d\tau$$

is a convolution of l in $\mathscr{C}([0,\varphi(b)])$ and hence, according to Theorem 1.3.6, the operation

$$(f*g)(x)=T^{-1}[(Tf)\,\tilde{*}\,(Tg)]=\int_0^{\varphi(x)} f\{\varphi^{-1}[\varphi(x)-\tau]\}g[\varphi^{-1}(\tau)]d\tau$$

is a convolution of L in $\mathscr{C}([a,b])$. By an easy transformation it can acquire the form

$$(f*g)(x)=\int_a^x f\{\varphi^{-1}[\varphi(x)-\varphi(\xi)]\}g(\xi)d\varphi(\xi). \tag{21}$$

The operator L can be represented in the form

$$Lf=\{1\}*f.$$

The convolutions of the form (21) had been studied systematically by Gesztelyi and Száz [60] and by Gesztelyi [59]. Earlier such a convolution had been used by Muravyev [81] for an operational calculus for an initial right inverse operator of the differential operator of the first order $D=\alpha(x)(d/dx)$, named by him "*α-operational calculus.*"

In using Theorem 1.3.6, sometimes a difficulty arises when we try to find the inverse operator T^{-1}. Sometimes T^{-1} can be replaced by another isomorphism $U:\tilde{\mathfrak{X}}\to\mathfrak{X}$.

Theorem 1.3.7. *Let $T:\mathfrak{X}\to\tilde{\mathfrak{X}}$ be a similarity of the linear operator $L:\mathfrak{X}\to\mathfrak{X}$ to a linear operator $\tilde{L}:\tilde{\mathfrak{X}}\to\tilde{\mathfrak{X}}$, and let the operation $\tilde{*}:\tilde{\mathfrak{X}}\times\tilde{\mathfrak{X}}\to\tilde{\mathfrak{X}}$ be an annihilators-free convolution of $\tilde{L}$ in $\tilde{\mathfrak{X}}$. If the operator $TU:\tilde{\mathfrak{X}}\to\tilde{\mathfrak{X}}$ is a multiplier of the convolution $\tilde{*}$ then the operation*

$$x*y=U[(Tx)\,\tilde{*}\,(Ty)] \tag{22}$$

is a convolution of L in $\mathfrak{X}$.

Proof. The bilinearity and commutativity of (22) follow from the corresponding properties of the convolution $\tilde{*}$ in $\tilde{\mathfrak{X}}$. For arbitrary $x,y,z\in\mathfrak{X}$ we have

$$\begin{aligned}(x*y)*z&=U[(TU(Tx\,\tilde{*}\,Ty))\,\tilde{*}\,Tz]\\&=UTU[(Tx\,\tilde{*}\,Ty)\,\tilde{*}\,Tz]=UTU[Tx\,\tilde{*}\,(Ty\,\tilde{*}\,Tz)]\\&=U[Tx\,\tilde{*}\,TU(Ty\,\tilde{*}\,Tz)]=x*(y*z).\end{aligned}$$

Here we have used twice the assumption that $\tilde{*}$ is a convolution of the operator TU.

Let us verify the convolutional property $L(x*y)=(Lx)*y$ of (22). Since by hypothesis TU and $\tilde{L}$ are multipliers of the annihilators-free convolutional algebra $(\tilde{\mathfrak{X}}, \tilde{*})$, then according to Theorem 1.3.1, the operators TU and $\tilde{L}$ commute, i. e. $\tilde{L}TU=TU\tilde{L}$. Then

$$\begin{aligned} L(x*y)&=(T^{-1}\tilde{L}\,T)U(Tx\,\tilde{*}\,Ty) \\ &=T^{-1}\tilde{L}\,(TU)\,(Tx\,\tilde{*}\,Ty)=T^{-1}(TU)\tilde{L}\,(Tx\,\tilde{*}\,Ty) \\ &=U(\tilde{L}\,Tx\,\tilde{*}\,Ty)=U(TLx\,\tilde{*}\,Ty)=(Lx)*y, \end{aligned}$$

where we have used the relation $TL=\tilde{L}T$, which expresses the similarity of L and $\tilde{L}$. □

The hypothesis of the theorem can be slightly strengthened in a way that the assumption that TU is a multiplier of the convolution $*$ to be not only sufficient, but a necessary condition in order (22) to be a convolution of L. One such strengthening is the requirement the convolutional algebra $(\tilde{\mathfrak{X}}, \tilde{*})$ to be factorable. This is so in the space of locally integrable functions for the Duhamel convolution, as we have seen in 1.1.4.

1.3.4. Convolutions of right inverse operators. Usually, the endomorphisms of linear spaces, for which we are seeking convolutions are right inverse operators.

Definition 5. Let $\mathfrak{X}$ be a linear space, $\mathfrak{X}_D$ its subspace, and let $D: \mathfrak{X}_D \to \mathfrak{X}$ be a linear operator. A linear operator $L: \mathfrak{X} \to \mathfrak{X}_D$ is said to be a *right inverse* of D, if $DLx=x$ for each $x \in \mathfrak{X}$.

Example. Let $\mathfrak{X}=\mathscr{C}([0, 1])$, and $D=d/dt$ with the domain $\mathfrak{X}_D =\mathscr{C}^1([0, 1])$. Then every right inverse of d/dt in $\mathscr{C}([0, 1])$ has the form

$$Lf(t)=\int_0^t f(\tau)d\tau+\Phi(f),$$

where $\Phi(f)$ is a linear functional.

With each right inverse operator a projector operator can be related This is the operator

$$F=[D, L]=I-LD \tag{23}$$

defined in $\mathfrak{X}_D$, where I is the identity of $\mathfrak{X}_D$. The operator F has the following properties:

1. The range of F coincides with ker D, i. e.

$$F(\mathfrak{X}_D)=\ker D.$$

2. The operator F is a left annihilator of L, i. e.

$$FL=0.$$

Indeed, $FL=(I-LD)L=L-LDL=L-L=0$.

3. The operator F is a projector, i. e. $F^2=F$. Indeed, $F^2=(I-LD) \cdot (I-LD)=I-2LD+LDLD=I-LD=F$.

In the above example we have

$$Ff(t)=f(0)-\Phi(f').$$

In the special case $\Phi(f)=0$, we get $Ff(t)=f(0)$.

Such projector operator F can sometimes be considered not only in the domain $\mathfrak{X}_D=\mathscr{C}^1([0,1])$, but in the whole space $\mathfrak{X}=\mathscr{C}([0,1])$ too, looking on 1, 2 and 3 as on defining properties of F.

Definition 6 (Przeworska-Rolewicz [85]). A linear operator $F: \mathfrak{X}_F \to \mathfrak{X} \supset \mathfrak{X}_F$ is said to be a defining projector of a right inverse operator L of an operator $D: \mathfrak{X}_D \to \mathfrak{X} \supset \mathfrak{X}_D$, provided it fulfils the conditions: (a) $\mathfrak{X}_D \subset \mathfrak{X}_F \subset \mathfrak{X}$; (b) $F(\mathfrak{X}_D)=\ker D$; (c) $FL=0$; and (d) $F^2=F$.

In the sense of this definition operator (23) is a defining projector of the right inverse operator L of D. We shall show that each defining projector of L is an extension of (23).

Lemma 3. *The restriction of an arbitrary defining projector of a right inverse operator L of $D: \mathfrak{X}_D \to \mathfrak{X}$ to $\mathfrak{X}_D$ coincides with the operator $I-LD$.*

Proof. Let $x \in \mathfrak{X}_D$. Denoting $LDx=y$, we have $Dx=Dy$, i. e. $x-y \in \ker D$. But $Fy=FDLx=0$ and since $F(x-y)=x-y$, then $Fx=x-y =x-LDx$. The lemma is proved. □

Corollary. If F is an arbitrary defining projector of a right inverse operator L of D, then the only common element of the kernels of D and F is the zero element, i. e.

$$\ker F \cap \ker D=\{0\}.$$

Indeed, if $x \in \ker F \cap \ker D$, then $Fx=0$ and $Dx=0$. Since, according to Lemma 3, $Fx=x-LDx$, then $x=0$.

The basic fact in the theory of the right-inverse operators, proposed by Przeworska-Rolewicz [85], is the *generalized Taylor formula*.

Lemma 4. (Bittner [3]). *If $L: \mathfrak{X}_D \to \mathfrak{X}$ is a linear right inverse of a linear operator $D: \mathfrak{X}_D \to \mathfrak{X} \supset \mathfrak{X}_D$, and if $F=I-LD$, then in $\mathfrak{X}_{D^n}$, the domain of D^n, the identity*

$$I=\sum_{k=0}^{n-1} L^k F D^k + L^n D^n \tag{24}$$

is fulfilled for $n=1, 2, \dots$. Here by I the identity of $\mathfrak{X}_{D^n}$ is denoted.

Proof. Formula (24) is only a short way for writing the obvious identity

$$I=(I-LD)+(LD-L^2D^2)+\cdots+(L^{n-1}D^{n-1}-L^nD^n)+L^nD^n.$$

Of course, instead of $F=I-LD$, we can take an arbitrary defining projector of L, since, according to Lemma 3, the restriction of such a projector to $\mathfrak{X}_D$ coincides with $I-LD$. □

Example. If $\mathfrak{X}=\mathscr{C}([0,1])$, $D=d/dt$, and L is the Volterra integration operator l, then $Ff(t)=f(t)-\int_0^t f'(\tau)d\tau=f(0)$ is defined not only in $\mathfrak{X}_D =\mathscr{C}^1([0,1])$, but in $\mathscr{C}([0,1])$ too. Then (24) is exactly the Maclaurin formula

$$f(t)=\sum_{k=0}^{n-1} f^{(k)}(0)\frac{x^k}{k!}+l^n f^{(n)}(t).$$

Now we shall specialize the notion of convolution of an endomorphism of linear space, introduced in 1.3.1 to the case of a right inverse operator. If $D: \mathfrak{X}_D \to \mathfrak{X} \supset \mathfrak{X}_D$ is a linear operator, then, by definition, each of its right inverse operators $L: \mathfrak{X} \to \mathfrak{X}_D$ is an endomorphism of the linear space $\mathfrak{X}$. According to the corollary of Lemma 3, the right inverse operator L of D is uniquely defined by D and F. Indeed, if y is the solution of the problem

$$Dy = x, \; Fy = 0,$$

then $z = y - Lx$ is a solution of the problem $Dz = 0$, $Fz = 0$, i. e. $z \in \ker D \cap \ker F$. But the only such solution is the zero element, and hence $y = Lx$. Therefore it is desirable to characterize the convolutions of L by means of D and F only. Under some further restrictions on the convolutions, this is done by the following theorem:

Theorem 1.3.8. *Let $\mathfrak{X}$ be a linear space, and let $L: \mathfrak{X} \to \mathfrak{X}$ be a right inverse of a linear operator $D: \mathfrak{X}_D \to \mathfrak{X} \supset \mathfrak{X}_D$. A bilinear, commutative and associative operation $*$ in $\mathfrak{X}$, such that $\mathfrak{X}_D$ is an ideal in the corresponding algebra, is a convolution of L in $\mathfrak{X}$ iff the following two conditions are fulfilled*: (a) *for each $x \in \mathfrak{X}_D$ and for each $y \in \mathfrak{X}$ the identity*

$$D(x * y) = (Dx) * y + D[(Fx) * y] \tag{25}$$

is fulfilled; (b) $\ker F$ *is an ideal in the algebra* $(\mathfrak{X}, *)$, *where* $F = I - LD$.

Proof. Let $x \in \mathfrak{X}_D$ and $y \in \mathfrak{X}$, and $*: \mathfrak{X} \times \mathfrak{X} \to \mathfrak{X}$ be a convolution of L in $\mathfrak{X}$, such that $\mathfrak{X}_D$ is an ideal in the algebra $(\mathfrak{X}, *)$. Then, from Lemma 3, $x = LDx + Fx$ and $x * y = L(Dx * y) + (Fx) * y$. By applying the operator D to both sides of this equality, we get (25). If $Fx = 0$, then $x = LDx$, and hence $x * y = L(Dx * y)$. Therefore, $F(x * y) = 0$, since $FL = 0$. Hence $\ker F$ is an ideal in the algebra $(\mathfrak{X}, *)$.

Conversely, if for each $x \in \mathfrak{X}_D$ and for each $y \in \mathfrak{X}$ the identity (25) is fulfilled, and $\ker F$ is an ideal in the algebra $(\mathfrak{X}, *)$, then if we substitute x in (25) by Lx, we get $D(Lx * y) = x * y$. But $Lx \in \ker F$ and hence $(Lx) * y \in \ker F$, i. e. $F[(Lx) * y] = 0$, since we have assumed that $\ker F$ is an ideal in $(\mathfrak{X}, *)$. The element $z = L(x * y) - (Lx) * y$ satisfies the equations $Dz = 0$ and $Fz = 0$. From the corollary of Lemma 3, it follows that $z = 0$, i. e. $L(x * y) = (Lx) * y$. Hence the operation $x * y$ is a convolution of L. □

Theorem 1.3.9. *Let $x * y$ be a convolution of the right inverse operator L of the right invertible operator $D: \mathfrak{X}_D \to \mathfrak{X} \supset \mathfrak{X}_D$, such that $\mathfrak{X}_D$ is an ideal in the corresponding convolutional algebra. Then, for all $x \in \mathfrak{X}_D$ and $y \in \mathfrak{X}$ the identity*

$$F(x * y) = F(Fx * y), \tag{26}$$

where F is an arbitrary defining projector of L, is fulfilled.

Proof. Since, according to Lemma 3, $Fx = x - LDx$ for $x \in \mathfrak{X}_D$, then $x = LDx + Fx$, and hence $x * y = L(Dx * y) + (Fx) * y$. Hence $F(x * y) = FL(Dx * y) + F[(Fx) * y] = F[(Fx) * y]$, since $FL = 0$. □

Corollary. If $x, y \in \mathfrak{X}_D$, then under the hypothesis of the theorem, the identity

$$F(x * y) = F[(Fx) * (Fy)]$$

(27) is fulfilled.

Example. Let $\mathfrak{X}=\mathscr{C}([0,1])$, $D=d/dt$, and let $L=l$ be the Volterra integration operator. If $f*g$ is the Duhamel convolution, then (25) takes the form

$$(f*g)'=f'*g+f(0)g$$

for $f\in\mathscr{C}^1([0,1])$ and $g\in\mathscr{C}([0,1])$.

Now we shall consider the problem for finding the multiplier operators of a convolution of right inverse operator. Very useful in some cases is a representation formula of the multipliers of such a convolution.

Theorem 1.3.10. *Let $*$ be a convolution of a right inverse operator $L\colon \mathfrak{X}\to\mathfrak{X}_D\subset\mathfrak{X}$ of a linear operator $D\colon \mathfrak{X}_D\to\mathfrak{X}$. If L has a representation as a convolutional operator*

$$(28)\qquad Lx=r*x,\ \text{with}\ r\in\mathfrak{X},$$

then each multiplier $M\colon \mathfrak{X}\to\mathfrak{X}$ of the convolutional algebra $(\mathfrak{X},*)$ has the form

$$(29)\qquad Mx=D(m*x),\quad m=Mr.$$

Proof. According to Theorem 1.3.1, the operator M commutes with L, i. e. $ML=LM$, since $(\mathfrak{X},*)$ is an annihilators-free algebra. The last assertion follows from the representation formula (28). Indeed, if $x*y=0$ for all $x\in X$, then $Ly=r*y=0$, and hence $y=0$.

Then, applying the operator M to (28), and using the commutativity of M and L, and the multiplier relation $M(x*y)=(Mx)*y$, we get

$$LMx=(Mr)*x.$$

Now, applying the operator D to this equality, we get the desired representation (29). □

Corollary 1. If $m=Mr$ is an element of $\mathfrak{X}_D$, and the convolution $*$ fulfils the hypothesis of Theorem 1.3.9, then representation (29) can be transformed into

$$(30)\qquad Mx=(Dm)*x+D[(Fm)*x].$$

Indeed, from Theorem 1.3.9 it follows

$$D(m*x)=(Dm)*x+D[(Fm)*x].$$

Corollary 2. Under the same hypothesis, as in the previous corollary, and under the additional assumption that if $x\in\ker D$, then $x*y\in\ker F$ for each $y\in\mathfrak{X}$, each operator of the form (30) is a multiplier of the convolutional algebra $(\mathfrak{X},*)$

In fact, the operator $M_1x=(Dm)*x$ is always a multiplier of $x*y$. We shall show that under the hypothesis made, the operator $M_2x=D[(Fm)*x]$ is a multiplier of $x*y$ too. Indeed

$$M_2(x*y)=D[(Fm)*(x*y)]=D[((Fm)*x)*y]$$
$$=D[(Fm)*x]*y+D[F((Fm)*x)*y]=(M_2x)*y$$

since $F[(Fm)*x]=0$, due to $Fm\in\ker D$.

1.3.5. Continuous convolutions of a Fréchet space endomorphism with a cyclic element. Let $\mathfrak{X}$ be a Fréchet space, and let $L\colon \mathfrak{X}\to\mathfrak{X}$ be an endomorphism. We assume that L has a cyclic element, i. e. an element $k \in \mathfrak{X}$, such that the set of all linear combinations of $\{L^n k\}$, $n=0, 1, 2, \ldots$, which we shall denote by span $\{L^n k\}_{n=0}^{\infty}$, is dense in $\mathfrak{X}$. The availability of a cyclic element of L in $\mathfrak{X}$ allows to prove some more subtle results for the continuous and annihilators-free convolutions of L in $\mathfrak{X}$.

Definition 7. A convolution $*$ in $\mathfrak{X}$ is said to be *separately continuous*, iff $x_n \to x$ implies $x_n * y \to x * y$ for each $y \in \mathfrak{X}$.

Sometimes, instead of separately continuous we shall say simply "continuous."

Theorem 1.3.11. *If* $*$ *is a separately continuous and annihilators-free convolution of an endomorphism* $L: \mathfrak{X} \to \mathfrak{X}$ *of a Fréchet space* $\mathfrak{X}$ *with a cyclic element, then the multiplier ring of the convolutional algebra* $(\mathfrak{X}, *)$ *coincides with the commutant of the operator* L *in* $\mathfrak{X}$.

Under the *commutant* of L in $\mathfrak{X}$ it is understood the set of all continuous endomorphisms $M: \mathfrak{X} \to \mathfrak{X}$ which commute with L.

Proof. Let $k \in \mathfrak{X}$ be a cyclic element of the operator L. We shall show that each endomorphism $M: \mathfrak{X} \to \mathfrak{X}$ (i. e. a continuous linear operator) with $ML=LM$ in $\mathfrak{X}$, is a multiplier of the convolution $*$. The argument in this case repeats the argument of 1.1.3 for the Duhamel convolution. From the commuting of M and L and from the assumption that $*$ is a convolution of L, it follows that for all non-negative integers m and n the obvious identity

$$(ML^m k) * (L^n k) = (L^m k) * (ML^n k)$$

is fulfilled. Due to the bilinearity of the operation $*$, we have proved the identities $Mx_m * y_n = x_m * My_n$ with arbitrary $x_m = \sum_{p=0}^{m} \alpha_{mp} L^p k$ and $y_n = \sum_{q=0}^{n} \beta_{nq} L^q k$, where α_{mp} and β_{nq} are constants.

Let $x, y \in \mathfrak{X}$ be arbitrarily chosen. From the assumption that k is a cyclic element of L, it follows that there exist sequences x_m and y_n of the above kind, such that $x_m \to x$, and $y_n \to y$ in $\mathfrak{X}$. Letting first $m \to \infty$, leaving n fixed, we get $Mx * y_n = x * My_n$. Now, letting $n \to \infty$, we get $Mx * y = x * My$. The complete multiplier relation $M(x * y) = (Mx) * y$ follows from Lemma 1, since the convolution $*$ is assumed to be annihilators-free.

According to Theorem 1.31, each multiplier M **of** $*$ **is a linear operator, commuting with** L**, i.e. it belongs to the commutant of** L**. As** 1.1 **the closed graph theorem implies continuity of M.** □

Corollary 1. All annihilators-free and continuous convolutions of an endomorphism of a *Fréchet space with a cyclic element* have the same multiplier ring.

In fact, this multiplier ring is the commutant of L in $\mathfrak{X}$.

Corollary 2. If $*$ is a continuous convolution in a Fréchet space $\mathfrak{X}$ of a linear right inverse operator $L: \mathfrak{X} \to \mathfrak{X}_D \subset \mathfrak{X}$ of a right-invertible operator $D: \mathfrak{X}_D \to \mathfrak{X}$ with a cyclic element in $\mathfrak{X}$, and with a convolutional representation $Lx = r * x$, then every continuous endomorphism $M: \mathfrak{X} \to \mathfrak{X}$, which commutes with L in $\mathfrak{X}$, admits a representation of the form

$$Mx = D(m * x), \quad m = Mr. \tag{31}$$

Remark. Theorem 1.3.11 is true for DF-spaces too, since its proof depends on the closed graph theorem.

Example 1. Let G be a domain in the complex plane, which is star-like with respect to the origin. Let $\mathscr{H}(G)$ be the space of all analytic functions in G. Let us find the commutant of the integration operator $lf(z) = \int_0^z f(\zeta)d\zeta$ in $\mathscr{H}(G)$.

In $\mathscr{H}(G)$ the Duhamel convolution

$$(f*g)(z)=\int_0^z f(z-\zeta)g(\zeta)d\zeta$$

is a bilinear, commutative and associative operation too, and $lf=\{1\}*f$. Here $\mathscr{H}(G)$ is a Fréchet space, and $f*g$ is a continuous convolution of l in $\mathscr{H}(G)$. The operator l has the constant function $\{1\}$ as a cyclic element, since, according to *Runge's theorem* (see Shabat [90], p. 124), the polynomials are dense in $\mathscr{H}(G)$. Then, using Corrollary 2, we get that each linear operator $M: \mathscr{H}(G)\to\mathscr{H}(G)$ which commutes with l in $\mathscr{H}(G)$ has a representation of the form

$$Mf(z)=\frac{d}{dz}\int_0^z m(z-\zeta)f(\zeta)d\zeta$$

with $m=M\{1\}$. After an easy transformation, we get

$$Mf(z)=m(0)f(z)+\int_0^z m'(z-\zeta)f(\zeta)d\zeta. \tag{32}$$

This representation is due to Raichinov [86] and it had been obtained in another way. It is easy to see that each operator of the form (32) commutes with l.

Now we give an example, showing that the assumption for availability of a cyclic element of L is an essential one for the validity of representation (31) of the commutant of L.

Example 2. Let $\mathfrak{X}=\mathscr{H}(G)$ be the space of the analytic functions in the unit disc $G=\{z\in\mathbb{C}:|z|<1\}$. Let us find the commutant of the operator l^2 in $\mathscr{H}(G)$.

Though the Duhamel convolution is a convolution of l^2 in $\mathscr{H}(G)$ too, and l^2 is representable as $l^2f=\{z\}*f$, formula (32) gives only a part of the operators $M: \mathscr{H}(G)\to\mathscr{H}(G)$ which commute with l^2. In order to make this clear, let us consider the splitting of $\mathscr{H}(G)$ as a direct sum of the spaces $\mathscr{H}_e(G)$ of the even functions and the space $\mathscr{H}_o(G)$ of the odd functions of $\mathscr{H}(G)$, i. e.

$$\mathscr{H}(G)=\mathscr{H}_e(G)\oplus\mathscr{H}_o(G).$$

The spaces $\mathscr{H}_e(G)$ and $\mathscr{H}_o(G)$ are invariant subspaces of $\mathscr{H}(G)$ for the operator l^2.

Let $M: \mathscr{H}(G)\to\mathscr{H}(G)$ be a linear operator from the commutant of l^2 in $\mathscr{H}(G)$, i. e. with $Ml^2=l^2M$. Let us find how the operator M acts on $\mathscr{H}_e(G)$ and $\mathscr{H}_o(G)$. The operator l^2 has the function $\{z\}$ as a cyclic element in $\mathscr{H}_o(G)$ and the function $\{1\}$ as a cyclic element in $\mathscr{H}_e(G)$. The Duhamel convolution

$$f*g(z)=\int_0^z f(z-\zeta)g(\zeta)d\zeta$$

is an inner operation in $\mathscr{H}_o(G)$, but not in $\mathscr{H}_e(G)$. But the operation

(33) $$f \overset{e}{*} g = l(f * g)$$

is as a convolution of l^2 in $\mathscr{H}_e(G)$, and an inner operation in $\mathscr{H}_e(G)$. Moreover, $l^2 f = \{1\} \overset{e}{*} f$. Since M and l^2 commute in $\mathscr{H}(G)$, then $Ml^2 f = l^2 Mf$ for each $f \in \mathscr{H}_e(G)$. By means of the same argument, as in the proof of Theorem 1.3.11, it is seen that the equality

(34) $$(Mf) \overset{e}{*} g = f \overset{e}{*} (Mg)$$

is an identity in $\mathscr{H}_e(G)$. If we substitute $g = \{1\}$, we get $(M\{1\}) \overset{e}{*} f = \{1\} \overset{e}{*} \{Mf\} = l^2 Mf$. Therefore

(35) $$Mf = \frac{d}{dz}(m * f)$$

with $m = M\{1\}$. Thus for even f we have a representation of the form (32).

Let now $f \in \mathscr{H}_o(G)$. As we have already mentioned, the Duhamel convolution $*$ is a convolution of l^2 in its invariant subspace $\mathscr{H}_o(G)$. Since the function $\{z\}$ is a cyclic element of l^2 in $\mathscr{H}_o(G)$, then, according to Theorem 1.3.11, M is a multiplier of the Duhamel convolution in $\mathscr{H}_o(G)$, i. e.

(36) $$(Mf) * g = f * (Mg)$$

for all $f, g \in \mathscr{H}_o(G)$. Substituting $g = \{z\}$ in (36), we get $(M\{z\}) * f = \{z\} * Mf = l^2 Mf$. Therefore

(37) $$Mf(z) = \frac{d^2}{dz^2}(n * f)$$

with $n = M\{z\}$.

Since each $f \in \mathscr{H}(G)$ is uniquely representable in the form $f = f_e + f_o$ with $f_e \in \mathscr{H}_e(G)$, $f_o \in \mathscr{H}_o(G)$ and $f_e(z) = \frac{f(z)+f(-z)}{2}$, $f_o(z) = \frac{f(z)-f(-z)}{2}$, then the general representation of a linear operator $M: \mathscr{H}(G) \to \mathscr{H}(G)$, commuting with l^2 in $\mathscr{H}(G)$ has the form

(38) $$Mf = (m * f_e)' + (n * f_o)''$$

with some functions $m, n \in \mathscr{H}(G)$.

Conversely, now we shall show that each operator of the form (38) with arbitrary m and n from $\mathscr{H}(G)$ commutes with l^2 in $\mathscr{H}(G)$. Indeed, from (38) we have $Ml^2 f = l(m * f_e) + n * f_o$. But applying l to (38), we get $lMf = m * f_e + (n * f_o)' - (n * f_o)'\,|_{z=0} = m * f_e + (n * f_e)'$, since $(n * f_e)'(0) = n(0) f_o(0) = 0$, having in mind that f_o is an analytic function. Then $l^2 Mf = l(m * f_e) + n * f_e$, and thus the relation $Ml^2 = l^2 M$ is proved.

Representation (38) contains two arbitrary functions m and n, and it can easily be seen that they cannot be reduced to one function, as in (32). Hence the commutants of l and l^2 in $\mathscr{H}(G)$ are different. The commutant of l is contained in the commutant of l^2, but there are operators $M: \mathscr{H}(G) \to \mathscr{H}(G)$ which commute with l^2, but they do not commute with l in $\mathscr{H}(G)$.

If we have a separately continuous and annihilators-free convolution $*$ of a right inverse operator $L: \mathfrak{X} \to \mathfrak{X}_D \subset \mathfrak{X}$ in a Fréchet space $\mathfrak{X}$ of an operator $D: \mathfrak{X}_D \to \mathfrak{X}$, then if L is a convolutional operator with a cyclic element, a representation formula can be shown, containing all possible continuous convolutions of L in $\mathfrak{X}$.

Theorem 1.3.12. *If $*$ is a separately continuous and annihilators-free convolution of a right inverse operator $L: \mathfrak{X} \to \mathfrak{X}_D \subset \mathfrak{X}$ in a Fréchet space $\mathfrak{X}$ with a representing element $r \in \mathfrak{X}$ of a right invertible operator $D: \mathfrak{X}_D \to \mathfrak{X}$, and if L has a cyclic element in $\mathfrak{X}$, then each separately continuous convolution $\tilde{*}$ of L in $\mathfrak{X}$ has a representation of the form*

$$x \,\tilde{*}\, y = D^2[\varrho * (x * y)] \tag{39}$$

with $\varrho = r \,\tilde{*}\, r$.

Proof. Let $\tilde{*}$ be an arbitrary separately continuous convolution of the operator $Lx = r * x$ in $\mathfrak{X}$. Then the given convolution $*$ is an annihilators-free convolution, since r is a non-zero non-divisor of zero of this convolution. The convolutional relation $L(x \,\tilde{*}\, y) = (Lx) \,\tilde{*}\, y$ expresses that the convolutional operator $M_x y = x \,\tilde{*}\, y$ commutes with L. Then, by Theorem 1.3.11, the operator M_x is a multiplier of the convolution $*$. In other words, we have shown that for every $x, y, z \in \mathfrak{X}$ the mixed associativity relation

$$x \,\tilde{*}\, (y * z) = (x \,\tilde{*}\, y) * z \tag{40}$$

is fulfilled. But representation formula (31) gives

$$\begin{aligned} x \,\tilde{*}\, y &= M_x v = D[(M_x r) * y] = D[(M_r x) * y] \\ &= D[(r \,\tilde{*}\, x) * y] = D[r \,\tilde{*}\, (x * y)], \end{aligned} \tag{41}$$

where we have used (40) too. But $r \,\tilde{*}\, (x * y) = M_r(x * y)$ and therefore

$$x \,\tilde{*}\, y = D[D[(r \,\tilde{*}\, r) * (x * y)]],$$

which is the desired representation (3). □

If $\varrho \in \mathfrak{X}$ is arbitrarily chosen, then one cannot assert that the operation $\tilde{*}$, defined by (39), is always a convolution of L in $\mathfrak{X}$. It may happen that (39) is not defined in the whole space $\mathfrak{X}$, but if this is not the case, then it may happen that (39) is non-associative. As for the convolution relation $L(f \,\tilde{*}\, g) = (Lf) \,\tilde{*}\, g$, from representation (39) follows only the weaker multiplier relation $(Lx) \,\tilde{*}\, y = x \,\tilde{*}\, (Ly)$. The complete multiplier relation follows only if we had ensured $\tilde{*}$ to be an associative and annihilators-free operation. A sufficient condition for the associativity of $\tilde{*}$, as defined by (39), is the requirement $F(x * y) = 0$ for all $x, y \in \mathfrak{X}$, where F is the defining projector of L. Then, in order to ensure $\tilde{*}$ to be annihilators-free, it is necessary to assume that $\varrho \neq 0$ is not a divisor of zero of $*$.

Theorem 1.3.13. *Under the hypothesis of Theorem 1.3.12, and assuming that the convolutional algebra $(\mathfrak{X}, *)$ is factorable, i. e. that each element $x \in \mathfrak{X}$ can be represented in the form $x = y * z$, with some $y, z \in \mathfrak{X}$, then each continuous convolution of L in $\mathfrak{X}$ admits a representation of the form*

(42) $$x \tilde{*} y = D[\varphi * (x * y)]$$
with $\varphi \in \mathfrak{X}$.

Proof. According to (41), we have $f \tilde{*} g = D[r \tilde{*} (f * g)]$. From the factorability of the algebra $(\mathfrak{X}, *)$, the operator

(43) $$Nx = D[r \tilde{*} x]$$

is defined in the whole space $\mathfrak{X}$. We shall show that N is a multiplier of the convolution $*$. If $x, y \in \mathfrak{X}$ are arbitrary, and if $x = x_1 * x_2$ is a factorization of x, then $N(x * y) = x_1 \tilde{*} (x_2 * y) = (x_1 \tilde{*} x_2) * y = (Nx) * y$, using (40). Hence N is a multiplier of the convolutional algebra $(\mathfrak{X}, *)$. Therefore, it admits a representation of the form $Nx = D[\varphi * x]$ with $\varphi = Nr$ (Theorem 1.3.10). Hence

$$x \tilde{*} y = N(x * y) = D[\varphi * (x * y)]$$

and the theorem is proved. □

A sufficient condition for the factorability of the convolutional algebra $(\mathfrak{X}, *)$ is to be a Banach algebra with bounded approximative identity.

1.4. THE MULTIPLIER QUOTIENTS RING OF AN ANNIHILATORS-FREE CONVOLUTIONAL ALGEBRA

In sect. 1.2 we have introduced the convolution quotient ring of the Duhamel convolution and named it the Mikusiński ring. An analogous construction is possible for a general convolution too, provided it has a non-empty set of non-trivial non-divisors of zero. But, as we had seen at the end of sect. 1.2, there an example of a Duhamel convolutional algebra can be given in which this is not the case, though the algebra itself is annihilators-free. In this case, instead of the convolution quotients ring, which does not exist, we may consider a corresponding multiplier quotients ring. Such a ring exists for every annihilators-free convolution. In the special case when the set of the non-trivial non-divizors of zero is non-empty, the corresponding convolution quotient ring is isomorphic with the multiplier quotient ring of the same convolution. But the advantages of the multiplier quotients ring become more tangible in the case of spaces of functions of several variables.

1.4.1. Definition of the multiplier quotient ring for an annihilators-free convolutional algebra. Let $\mathfrak{X}$ be a linear space with an endomorphism $L: \mathfrak{X} \to \mathfrak{X}$ in it. We assume that L is a regular operator, i. e. that $Lx = 0$ implies $x = 0$. In other words, we suppose that $\lambda = 0$ is not an eigenvalue of the operator L.

But our main assumption is that the operator L has a convolution $*: \mathfrak{X} \times \mathfrak{X} \to \mathfrak{X}$ in $\mathfrak{X}$, i. e. that there is a bilinear commutative and associative operation $*: \mathfrak{X} \times \mathfrak{X} \to \mathfrak{X}$, such that L is a multiplier of the algebra $(\mathfrak{X}, *)$, i. e. that the relation

(1) $$L(x * y) = (Lx) * y$$

is fulfilled for all $x, y \in \mathfrak{X}$. Moreover, we should assume that the convolution $*$ is annihilators-free. The last assumption means that $x * y = 0$ for all $y \in \mathfrak{X}$ implies $x = 0$.

Let $\mathfrak{M}$ be the multiplier ring of the annihilators-free convolutional algebra $(\mathfrak{X}, *)$. By $\mathfrak{M}_*$ we shall denote the multiplicative subset of $\mathfrak{M}$, consisting of all non-trivial non-divisors of zero in $\mathfrak{M}$ with respect to the operator multiplication, i. e.

(2) $$\mathfrak{M}_*=\{M: M\in\mathfrak{M}, MN=0, N\in\mathfrak{M}\Rightarrow N=0\}.$$

The set $\mathfrak{M}_*$ is non-empty, since always $I\in\mathfrak{M}_*$ and $L\in\mathfrak{M}_*$, the last by assumption.

According to Theorem 1.3.1, the multiplier ring $\mathfrak{M}$ is a commutative one. Theorem 1.3.2 shows that the annihilators-free algebra $\mathfrak{A}=(\mathfrak{X}, *)$ can isomorphically be embedded into its multiplier ring $\mathfrak{M}$ by means of the natural embedding

(3) $$\xi: x\rightarrow x*,$$

where by $x*$ we denote the convolutional operator $(x*)y\overset{\text{def}}{=}x*y$ on $\mathfrak{X}$. The field $\mathbf{C}$ of the complex numbers can also be embedded isomorphically into $\mathfrak{M}$ by means of the map

(4) $$\eta: \lambda\rightarrow[\lambda],$$

where $[\lambda]$ denotes the numerical operator $[\lambda]x\overset{\text{def}}{=}\lambda x$. The embeddings ξ and η are compatible in the sense that

(5) $$\xi(\lambda x)=\eta(\lambda)\xi(x)=[\lambda](x*).$$

Here a remark should be made. We are speaking about an embedding of the convolutional algebra $\mathfrak{A}=(\mathfrak{X}, *)$ into the multiplier ring $\mathfrak{M}$ in the sense that $\xi(\mathfrak{A})$ is an algebra over the field $\eta(\mathbf{C})$. The "forgetting" of the scalar multiplication operation is intended to simplify the considerations. Indeed, we are to deal with the ring system $\mathfrak{M}$ only, a system with two operations only, instead of the three operations in the corresponding convolutional algebra. But the scalar multiplication operation is not forgotten at all, due to the fact that $\mathfrak{A}$ contains subrings isomorphic to the complex number field and to the convolutional ring of the algebra $(\mathfrak{X}, *)$. According to (5), the scalar multiplication in $\mathfrak{A}$ has a counterpart in $\mathfrak{M}$. It should not be forgotten that this is possible under the assumption that the convolutional algebra $(\mathfrak{X}, *)$ is annihilators-free.

The notations can be simplified in such a way as to cause no confusion. Since $\xi(\lambda x)=\eta(\lambda)\xi(x)$, then identifying $\xi(x)=x*=x$ we can write only x, and instead of the denotation $x*y$ for the convolutional product of x and y we write simply xy, regarding $\mathfrak{X}$ as a subring of $\mathfrak{M}$. The numerical operators $\eta(\lambda)=[\lambda]$ will be denoted only by λ. Thus, it is appropriate to denote the identity of $\mathfrak{M}$ by 1 instead of I, as we have done till now.

Definition 1. The *multiplier quotients ring* of an annihilators-free convolutional algebra $\mathfrak{A}=(\mathfrak{X}, *)$ is said to be the quotient ring $\mathfrak{M}_*^{-1}\mathfrak{M}$, defined by the factorization of the product set $\mathfrak{B}=\mathfrak{M}\times\mathfrak{M}_*$ with respect to the equivalence relation

(6) $$\mathfrak{R}=\{((M, N), (P, Q))\in\mathfrak{B}\times\mathfrak{B}: MQ=NP\}$$

with the usual operations in a quotient ring.

We shall denote

$$\mathsf{M}=\mathfrak{B}/\mathfrak{R}=\mathfrak{M}_*^{-1}\mathfrak{M}. \tag{7}$$

The elements of the multiplier quotient ring M can be denoted as usual fractions M/N with $M\in\mathfrak{M}$ and $N\in\mathfrak{M}_*$.

The addition and the multiplication in M are defined as usual by

$$\frac{M}{N}+\frac{P}{Q}=\frac{MQ+NP}{NQ} \tag{8}$$

and

$$\frac{M}{N}\frac{P}{Q}=\frac{MP}{NQ}. \tag{9}$$

Lemma 1. The map $\zeta:\mathfrak{M}\to\mathsf{M}$, defined by

$$\zeta: M\mapsto\frac{M}{1} \tag{10}$$

is an isomorphic embedding of $\mathfrak{M}$ in M.

The proof is evident. □

Since by means of maps (3) and (4) the convolutional algebra $\mathfrak{A}=(\mathfrak{X},*)$ and the complex number field are already embedded in $\mathfrak{M}$, then by the mappings $\zeta\xi$ and $\zeta\eta$ embeddings of $\mathfrak{X}$ and C in M are accomplished. More precisely, we have

$$\zeta\xi:\ x\mapsto\frac{x*}{[1]} \tag{11}$$

and

$$\zeta\eta:\ \lambda\mapsto\frac{[\lambda]}{[1]}. \tag{12}$$

If $M\in\mathfrak{M}$, we shall identify $M/1$ with M and write $M=M/1$, which would cause no confusion. In particular, we shall consider L as element of M.

Definition 2. The algebraic inverse of L is said to be the element S of the multiplier quotient ring M, which is reciprocal to L in $\mathfrak{M}$, i. e.

$$S=\frac{1}{L}. \tag{13}$$

We have $SL=LS=1$.

For many applications it is important to know when a polynomial expression of S is invertible in the ring M.

Theorem 1.4.1. *If $P(\lambda)=\sum_{k=0}^{n}\alpha_k\lambda^k$ is a polynomial of n-th degree with complex coefficients, then the element $P(S)$ of M is algebraically invertible in M iff any of the zeros of the reciprocal polynomial $P^*(\lambda)=\lambda^nP(1/\lambda)$ is not an eigenvalue of the operator L.*

Proof. Due to the algebraic closedness of C, the polynomial $P(\lambda)$ has a representation of the form

$$P(\lambda)=a_n\prod_{k=1}^{m}(\lambda-\lambda_k)^{\mu_k}, \qquad a_n\neq 0,$$

where $\lambda_k\in\mathrm{C}$, $k=1, 2, \ldots, m$, are its pairwise different zeros, and μ_k are their corresponding multiplicities. By assumption, $a_n\neq 0$. The product $P(S)=a_n\prod_{k=1}^{m}(S-\lambda_k)^{\mu_k}$ is invertible in M iff none of the elements $S-\lambda_k$ is a divisor of zero in M. That is why we shall answer the question: when is an element of the form $S-\lambda$ with $\lambda\in\mathrm{C}$ a divisor of zero in M? If $S-\lambda$ is a divisor of zero, then there exists a multiplier fraction A/B with $A\neq 0$, such that $(S-\lambda)A/B=0$, i. e. with $(S-\lambda)A=0$. The assumption $A\neq 0$ means that there is an element $a\in\mathfrak{X}$, such that $Aa\neq 0$. If we denote $Aa=u$, we have $(S-\lambda)u=0$. Multiplying this equality by L, we get $u-\lambda Lu=0$. Since $u\neq 0$, then $\lambda\neq 0$. Hence $1/\lambda$ is an eigenvalue of the operator L.

Conversely, if $\lambda\in\mathrm{C}$, $\lambda\neq 0$, and $1/\lambda$ is an eigenvalue of the operator L, then let $u\in\mathfrak{X}$, $u\neq 0$ be a corresponding eigenelement of L. Then we have $(1-\lambda L)u=0$. Multiplying by S, we get $(S-\lambda)u=0$. Hence $S-\lambda$ is a divisor of zero in M. Thus all is proved, since the zeros of the polynomial $P^*(\lambda)=\lambda^n P(1/\lambda)$ are the reciprocal values of the different zeros of $P(\lambda)$, and possibly $\lambda=0$. □

Corollary. If the polynomial expression $P(S)$ is invertible, then its reciprocal element $1/P(S)$ admits a representation of the form

$$\frac{1}{P(S)}=\sum_{k=1}^{m}\sum_{j=1}^{\mu_k}\frac{\beta_{k,j}}{(S-\lambda_k)^j}, \qquad \beta_{k,j}\in\mathrm{C}, \tag{14}$$

where in the right-hand side the standard partial fractions development of $1/P(\lambda)$ is used.

1.4.2. The ring of the convolution quotients of a convolution with non-divisors of zero. In sect. 1.2 we have introduced the Mikusiński ring, i. e. the convolution quotients ring of the Duhamel convolution. A similar construction for a general convolution in a linear space $\mathfrak{X}$ with a convolution with non-empty set of non-trivial non-divisors of zero is always possible.

Definition 3. Let $\mathfrak{A}=(\mathfrak{X}, *)$ be a convolutional algebra with non-empty set $\mathfrak{X}_*$ of non-trivial non-divisors of zero. The *convolution quotient ring* of $\mathfrak{A}$ is said to be the factor set of the product set $\mathfrak{B}=\mathfrak{X}\times\mathfrak{X}_*$ with respect to the equivalence relation

$$\mathfrak{R}=\{((x, y), (u, v))\in\mathfrak{B}\times\mathfrak{B}: x*v=y*u\} \tag{15}$$

in $\mathfrak{B}$ with the usual definitions for addition and multiplication of fractions.

Let us denote by $\mathfrak{N}$ the corresponding convolution quotients ring and its elements, provisionally, by $x:y$. The last part of the definition should be understood in the sense that

$$(a:b)+(c:d)=(a*d+b*c):(b*d) \tag{16}$$

and

$$(a:b).(c:d)=(a*c):(b*d). \tag{17}$$

We drop the trivial and well-known proof of the correctness of these definitions. The convolution quotients just introduced form a commutative and associative ring with a unit, and the map

$$x \mapsto (a*x):a \tag{18}$$

is an isomorphic embedding of the convolutional algebra $\mathfrak{A}=(\mathfrak{X}, *)$ into $\mathfrak{R}$, provided a is a non-trivial non-divisor of zero of the convolution considered.

Theorem 1.4.2. *If $\mathfrak{A}=(\mathfrak{X}, *)$ is a convolutional algebra with non-empty set $\mathfrak{X}_*$ of the non-trivial non-divisors of zero, then the convolution quotients ring $\mathfrak{R}$ of $\mathfrak{A}$ is isomorphic with the multiplier quotients ring* M *of $\mathfrak{A}$.*

Proof. Let $a\in\mathfrak{X}$, $a\neq 0$ be a non-divisor of zero of the algebra $\mathfrak{A}$. We shall show that the map

$$\sigma: x:y \mapsto \frac{x*}{y*} \tag{19}$$

of the convolution quotients ring $\mathfrak{R}$ into the multiplier quotients ring M is a ring isomorphism. Here $x*$ and $y*$ denote the multipliers $(x*)u=x*u$ and $(y*)u=y*u$, correspondingly.

First, we shall show that this map is well defined. Let $x:y=u:v$. Then $x*v=y*u$, and hence for each $z\in\mathfrak{X}$ we have $(x*v)*z=(y*u)*z$ or $x*(v*z)=y*(u*z)$. The last equality means that $(x*)(v*)=(y*)(u*)$, and hence $x*/y*=u*/v*$.

Next we show that each element M/N of M is an image of an element of $\mathfrak{R}$ under map (19). Indeed,

$$\sigma: (Ma):(Na) \mapsto \frac{M}{N}.$$

Now we shall show that the map (19) is one-to-one. In fact, let $x:y\neq u:v$. Then the images $x*/y*$ and $u*/v*$ of $x:y$ and $u:v$ are different too. Indeed, if $x*/y*=u*/v*$, then $(x*)(v*)=(y*)(u*)$ and $x*(v*z)=y*(u*z)$ for each $z\in\mathfrak{X}$. In particular, for $z=a$ we have $(x*v-y*u)*a=0$, but since a is a non-trivial non-divisor of zero, then $x*v-y*u=0$ in contradiction to the assumption $x:y\neq u:v$.

The fact that (19) is a ring isomorphism can be verified in a trivial manner. □

Therefore, when the convolution quotients ring exists, it is not essentially different from the multiplier quotients ring of the same convolution. But one can point out some quite natural examples of convolutions, in which there are no non-trivial non-divisors of zero, though the convolutions are annihilators-free. In such cases we can use only the construction of the multiplier quotient ring. It seems that Máté [75] was the first to point out this. In this paper the idea of introducing the multiplier quotient ring for the Duhamel convolution can be seen. In Struble [92] traces of such a construction can also be seen. In Dimovski [40] the scheme presented here is developed. Száz [94] had proposed a slightly more general scheme.

For brevity sake we shall speak of quotient ring M of a convolutional annihilators-free algebra $\mathfrak{A}=(\mathfrak{X}, *)$, and from the context it would be clear whether the convolution quotients, or multiplier quotients ring is meant.

Example. In a sense, the problem for construction of the quotient ring of an annihilators-free convolution in a linear space is a generalization of the problem for *adjoining a root* of an irreducible polynomial over a field. For the sake of simplicity, let us take the rational numbers field Q and let $P(\lambda)=\sum_{k=0}^{n} a_k\lambda^k$ with $a_n\neq 0$ be an irreducible polynomial on Q. The field extension $Q[\theta]$, obtained by adjoining of a root θ of $P(\lambda)=0$ to Q can be interpreted as a quotients ring of a suitable convolution in the space Q^n of the rational vectors $x=(x_1,\dots,x_n)^T$. Let us consider the matrix operator

$$L=\begin{Vmatrix} -\frac{a_0}{a_n} & 1 & 0 & \dots & 0 \\ -\frac{a_1}{a_n} & 0 & 1 & \dots & 0 \\ \dots & \dots & \dots & \dots & \dots \\ -\frac{a_{n-2}}{a_n} & 0 & 0 & \dots & 1 \\ -\frac{a_{n-1}}{a_n} & 0 & 0 & \dots & 0 \end{Vmatrix}. \tag{20}$$

The characteristic polynomial $\det(L-\lambda I)$ of L coincides with $P(\lambda)$ up to the multiplier $(-1)^n$, i.e. $\det(L-\lambda I)=(-1)^n P(\lambda)$. Since, on assumption, $P(\lambda)$ is irreducible on Q, then the order of an arbitrary non-zero vector x with respect to L is equal exactly to n (see [62], p. 90). Otherwise, for each $x\in Q$, $x\neq 0$, the vectors $x, Lx, L^2x,\dots, L^{n-1}x$ form a basis in Q^n. Let us take $x=e=(0, 0,\dots, 1)^T$. Then $Le=(0, 0,\dots, 1, 0)^T,\dots, L^{n-2}e =(0, 1,\dots, 0)^T$ and $L^{n-1}e=(1, 0,\dots, 0, 0)^T$. Since each vector $x=(x_1, x_2,\dots, x_n)^T$ of Q^n can be represented in the form

$$x=\sum_{k=1}^{n} x_k L^{n-k}e,$$

in order to define a convolution of L in Q^n it is sufficient to give a multiplication table $(L^pe)*(L^qe)$, $p, q=0, 1, 2,\dots, n-1$. It is most natural to assume

$$(L^pe)*(L^qe)=L^{p+q}e, \quad p, q=0, 1,\dots, n-1. \tag{21}$$

Thus we have taken e as the unit of the corresponding convolutional algebra. According to the Hamilton-Caley theorem, we have $P(L)=0$, and the powers of L with degrees $\geqq n$ should be replaced by sums of linear combinations of lower degrees. Defined in such a way, the operation $*$ is bilinear, commutative and associative. The operator L itself can be represented as the convolutional operator $(Le)*$, i.e. $Lx=(Le)*x$.

Hence, the operation $*$ is a convolution of L in Q^n. We shall show that $*$ is a convolution without divisors of zero in Q^n. Indeed, let $a\neq 0$,

$a \in \mathrm{Q}$ be such that $a * x = 0$ for some $x \in \mathrm{Q}^n$. Then $(L^k a) * x = 0$ for $k = 0, 1, 2, \ldots, n-1$. But, as we have already mentioned, the vectors $\{L^k a\}_{k=0}^{n-1}$ form a basis in Q^n and hence the vector e can be represented in the form

$$e = \sum_{k=0}^{n-1} \delta_k L^k a.$$

Hence $e * x = 0$, i. e. $x = 0$ since e is the unit of the convolutional algebra. Thus we have proved that $*$ is a convolution without divisors of zero.

The space Q^n with the operations $*$ and $+$ is a field, and hence it coincides with the quotient ring M of Q^n. Thus we have seen that the multiplier quotients ring construction is a generalization of a very common algebraic construction.

1.4.3. Isomorphism of the multiplier quotients rings of similar operators. In 1.3.3 we have shown a way of finding a convolution of a linear operator $L: \mathfrak{X} \to \mathfrak{X}$, when a convolution of an operator $\tilde{L}: \tilde{\mathfrak{X}} \to \tilde{\mathfrak{X}}$, similar to it, is known. If $T: \mathfrak{X} \to \tilde{\mathfrak{X}}$ is the similarity of L to $\tilde{L}$, and $\tilde{*}$ is a convolution of $\tilde{L}$ in $\tilde{\mathfrak{X}}$, then the operation

$$x * y = T^{-1}[(Tx) \tilde{*} (Ty)] \tag{22}$$

is a convolution of L in $\mathfrak{X}$. The corresponding convolutional algebras $(\mathfrak{X}, *)$ and $(\tilde{\mathfrak{X}}, \tilde{*})$ are isomorphic.

Lemma 1. *The map $\varphi: M \mapsto \tilde{M} = TMT^{-1}$ from the multipliers ring $\mathfrak{M}$ of $(\mathfrak{X}, *)$ to the multipliers ring $\tilde{\mathfrak{M}}$ of $(\tilde{\mathfrak{X}}, \tilde{*})$, with $*$ and $\tilde{*}$, connected as in (22), is a ring isomorphism.*

Proof. Let M be a multiplier of the convolution $*$, i. e. $M(x * y) = (Mx) * y$ for every $x, y \in \mathfrak{X}$. According to (22), this means that $MT^{-1}[(Tx) \tilde{*} (Ty)] = T^{-1}[(TMx) \tilde{*} (Ty)]$, which, written in the form $(TMT^{-1})[(Tx) \tilde{*} (Ty)] = [(TMT)^{-1}(Tx)] \tilde{*} (Ty)$ means that $\tilde{*}$ is a convolution of the operator $\tilde{M} = TMT^{-1}$. It is also clear that each multiplier $\tilde{M}$ of $\tilde{*}$ is an image of the multiplier $M = T^{-1}\tilde{M}T$ of $*$ under the map φ. It is obvious that

$$\varphi(M + N) = \varphi(M) + \varphi(N),$$

$$\varphi(MN) = TMNT^{-1} = (TMT^{-1})(TNT^{-1}) = \varphi(M)\varphi(N).$$

Hence, φ is a ring isomorphism. □

Corollary. If $M = m *$, then $\tilde{M} = (Tm) \tilde{*}$.

Indeed, if $Mx = m * x$, then $\tilde{M}\tilde{x} = TMT^{-1}\tilde{x} = T[m * T^{-1}\tilde{x}] = (Tm) \tilde{*} \tilde{x}$ and therefore $\tilde{M}\tilde{x} = (Tm) \tilde{*} \tilde{x}$ for each $\tilde{x} \in \tilde{\mathfrak{X}}$.

The isomorphism $\varphi: \mathfrak{M} \to \tilde{\mathfrak{M}}$ of the multiplier rings of the convolutional algebras $(\mathfrak{X}, *)$ and $(\tilde{\mathfrak{X}}, \tilde{*})$ connected by (22) can be extended to isomorphism of their corresponding multiplier quotients rings $\mathrm{M} = \mathfrak{M}_*^{-1}\mathfrak{M}$ and $\tilde{\mathrm{M}} = \tilde{\mathfrak{M}}_{\tilde{*}}^{-1} \tilde{\mathfrak{M}}$. In fact, the map

$$\psi: \frac{M}{N} \mapsto \frac{\varphi(M)}{\varphi(N)}, \tag{23}$$

where $\varphi(M)=TMT^{-1}$ and $\varphi(N)=TNT^{-1}$ is a ring isomorphism and its restriction to $\mathfrak{M}$ coincides with φ. That is why we may retain the same notation φ for ψ.

Example. In 1.3.3 we have considered as an example the operator

$$Lf(x)=\int_a^x f(\xi)d\varphi(\xi) \tag{24}$$

in $\mathfrak{X}=\mathscr{C}([a, b])$, where $\varphi(x)\in\mathscr{C}([a, b])$ with $\varphi(a)=0$ had been supposed to be strictly increasing. There we have shown that the map

$$T: f(x) \leftrightarrow \tilde{f}(t)=f[\varphi^{-1}(t)] \tag{25}$$

is a similarity from L to the Volterra integration operator

$$l\tilde{f}(t)=\int_0^t \tilde{f}(\tau)d\tau \quad \text{in} \quad \mathscr{C}([0, \varphi(b)]), \tag{26}$$

i. e. $TL=lT$. We have shown that the operation

$$(f*g)(x)=\int_a^x f[\varphi^{-1}(\varphi(x)-\varphi(\xi))]\, g(\xi)\, d\varphi(\xi) \tag{27}$$

is a convolution of L, corresponding to the Duhamel convolution by the map (25).

Let us consider the multiplier quotients ring for the Duhamel convolution in $\mathscr{C}([0, \varphi(b)])$ and the multiplier quotient ring for (27) in $\mathscr{C}([a, b])$. Let us denote by s the algebraic inverse of l, and by S the algebraic inverse of L, i. e. $s=1/l$ and $S=1/L$. From the results of sect. 1.2 we know that for an arbitrary $\alpha\in\mathbb{C}$ we have the representation

$$\frac{1}{(s-\alpha)^n}=\left\{\frac{t^{n-1}}{(n-1)!}e^{\alpha t}\right\}\tilde{*}, \quad n=1, 2, \ldots, \tag{28}$$

where the star $\tilde{*}$ means that the function is the kernel of a corresponding convolutional operator (with respect to the Duhamel convolution $\tilde{*}$). Then by the corollary of Lemma 2, we have

$$\frac{1}{(S-\alpha)^n}=\left\{\frac{[\varphi(x)]^{n-1}}{(n-1)!}e^{\alpha\varphi(x)}\right\}* \tag{29}$$

since $T^{-1}\tilde{f}(t)=\tilde{f}[\varphi(x)]$.

This example illustrates how to use in practice the similarity method. The use of the similarity method in this example is due to Gesztelyi and Száz [60], but this example had been considered earlier by Muravyev [81] directly.

1.4.4. Convolutional approach to Taylor boundary value problems for abstract differential equations. In 1.3.4 we presented the elements of the theory of the right inverse operators. In particular, we have found a characteristic formula for an annihilators-free convolution of a right inverse operator by means of the operator of which it is a right inverse and by means of its defining projector. In practice, since we mostly have to construct operational calculi for right inverse operators, it is desirable to specialize the general scheme for solving Taylor boundary value problems for abstract differential equations of a type connected with a right inverse operator.

Let $L: \mathfrak{X} \to \mathfrak{X}_D \subset \mathfrak{X}$ be a right inverse operator of a linear operator $D: \mathfrak{X}_D \to \mathfrak{X}$ and let F be its defining projector. Let us remind that the restriction of F to $\mathfrak{X}_D$ coincides with the commutator $[D, L] = I - LD$. If $P(\lambda) = \sum_{k=0}^{n} a_k \lambda^k$ is an n-th degree polynomial with complex coefficients, then by means of the polynomial $P(\lambda)$ we can introduce the n-th order *linear abstract differential equation*

$$P(D)x = f \tag{30}$$

with a given $f \in \mathfrak{X}$ and with the indeterminate x, which is sought in $\mathfrak{X}_{D^n}$, the domain of D^n.

Definition 4. A *Taylor boundary value problem* for the abstract differential equation (30), determined by the projector $F = I - LD$, is named the boundary value problem

$$FD^k x = \gamma_k, \quad k = 0, 1, \ldots, n-1, \tag{31}$$

with given $\gamma_k \in \ker D$, $k = 0, 1, \ldots, n-1$.

The class of the Taylor boundary value problems for abstract differential equations had been considered by Bittner [4], Przeworska-Rolewicz [85], Berg [1], and by Tasche [97]. Here we shall consider this problem in the special case when the right inverse operator L of D, defined by the projector F, has an annihilators-free convolution $*$ in $\mathfrak{X}$. Moreover, we impose a further restriction, namely, that the operator L is a convolutional operator in this convolution, i. e. there exists an element r, such that

$$Lx = r * x \quad \text{for all} \quad r \in \mathfrak{X}. \tag{32}$$

If we have at our disposal such a convolution, we can give an explicit representation of the remainder term in the generalized Taylor formula (24) of L from 1.3.3. We have shown that for each $x \in \mathfrak{X}_{D^N}$ the identity

$$x = Fx + LFDx + L^2FD^2x + \ldots + L^{N-1}FD^{N-1}x + L^N D^N x \tag{33}$$

holds.

Here $R_N(x) = L^N D^N x$ is the remainder term. Having in mind (32), if we denote $B_{N-1} = L^{N-1} r$, we get

$$x = \sum_{k=0}^{N-1} L^k FD^k x + B_{N-1} * D^N f. \tag{33'}$$

Example. As an illustration, let us consider the classical Taylor-Maclaurin formula

$$f(t)=f(0)+f'(0)\frac{t}{1!}+\ldots+f^{(N-1)}(0)\frac{t^{N-1}}{(N-1)!}+l^N f^{(N)}(t).$$

Since $lf=\{1\}*f$, then (33) gives

$$f(t)=\sum_{k=0}^{N-1} f^{(k)}(0)\frac{t^k}{k!}+\frac{1}{(n-1)!}\int_0^t (t-\tau)^{N-1} f^{(N)}(\tau)d\tau,$$

which is the same formula with Cauchy remainder term.

In the next chapters we shall show that some other well-known formulas from the classical analysis are in fact generalized Taylor formulas for some right inverse operators of the differentiation operator or its square. One may mention Euler-Maclaurin's formula, Boole's formula, Lidstone's formula. Here we shall find simple formulas for the remainder terms of such formulas.

If $*$ is a convolution of a right inverse operator L of a right invertible operator D, and L can be represented in this convolution as a convolutional operator, then the convolution considered is surely annihilators-free, since the representing element of L is necessarily a non-zero non-divizor of zero. In such a case we may restrict our attention to the convolution quotients ring, instead of the multiplier quotients ring, which is isomorphic to it.

Let $\mathfrak{X}_*$ be the set of the non-trivial non-divisors of zero of the convolution $*$, and let $\mathfrak{R}=\mathfrak{X}_*^{-1}\mathfrak{X}$. In $\mathfrak{R}$ we can write $L=r$, and let us denote by S the inverse element of L, i. e.

$$S=\frac{1}{L}. \tag{34}$$

Now the fraction $1/L$ is understood as an element of the convolution quotients ring $\mathfrak{R}$.

The generalized Taylor formula (33) can be written in the form

$$D^N x=S^N x-\sum_{k=0}^{N-1} S^{N-k}(FD^k x), \tag{35}$$

where $x\in\mathfrak{X}_{D^N}$ is arbitrarily chosen. It can be obtained from (33) by multiplying with S^N. By means of (35) each Taylor boundary value problem (30), (31) can be reduced to the following algebraic equation

$$P(S)x=f+\sum_{k=0}^{n-1}\left(\sum_{j=1}^{n-k} \alpha_{k+j}S^j\right)\gamma_k \tag{36}$$

in the convolution quotients ring $\mathfrak{R}$. Here $\alpha_0, \alpha_1, \ldots, \alpha_n$ are the coefficients of the polynomial $P(\lambda)=\sum_{k=0}^{n}\alpha_k\lambda^k$.

According to Theorem 1.4.1, a sufficient condition of the solvability of (36) in $\mathfrak{R}$ is the zeros of the polynomial $P^*(\lambda)=\lambda^n P(1/\lambda)$ not to be eigenvalues of the corresponding right inverse operator L. Then, we can represent equation (36) in the form

$$P(S)x=f+\sum_{k=0}^{n-1} P_k(S)y_k \tag{37}$$

with $P_k(S)=\sum_{j=1}^{n-k} a_{k+j}S^j$, $k=0, 1,\ldots, n-1$.

Theorem 1.4.3. *Let none of the zeros $\lambda_1, \lambda_2, \ldots, \lambda_n$ of the polynomial $P(\lambda)$ be an eigenvalue of the eigenvalue problem $Du-\lambda u=0$, $Fu=0$. If each of the partial fractions $1/(S-\lambda_k)$ is an element of $\mathfrak{X}$, then*

$$x=\frac{1}{P(S)}f+\sum_{k=0}^{n-1}\frac{P_k(S)}{P(S)}y_k \tag{38}$$

is a true solution of the Taylor boundary value problem (30), (31).

Proof. We shall show that if $g\in\mathfrak{X}_{D^N}$ for a positive integer N, then $y_k=g/(S-\lambda_k)\in\mathfrak{X}_{D^{N+1}}$. Indeed, since $1/(S-\lambda_k)\in\mathfrak{X}$, then $y_k\in\mathfrak{X}$. This y_k satisfies the equation $y_k=\lambda_k Ly_k+Lg$. If we assume $y_k\in\mathfrak{X}_{D^j}$ with $0\leq j<N+1$, from this equation it follows immediately that $y_k\in\mathfrak{X}_{D^{j+1}}$. Hence $y_k\in\mathfrak{X}_{D^{N+1}}$.

Since $f\in\mathfrak{X}$, then $[1/P(S)]f=\frac{1}{a_n}\prod_{k=1}^{n}(S-\lambda_k)^{-1}f\in\mathfrak{X}_{D^n}$ is an element of $\mathfrak{X}_{D^n}$. But from $y_k\in\ker D$, $k=0,1,\ldots, n-1$, it follows easily that $[P_k(s)/P(S)]y_k\in\mathfrak{X}_{D^n}$. Hence x, defined by (38), is an element of $\mathfrak{X}_{D^n}$. The verification that x is a solution of (30), (31) is done in converse order.

Finally, let us consider the special case of (30), (31) with zero boundary value conditions

$$P(D)u=f, \quad FD^k u=0, \quad k=0, 1,\ldots, n-1. \tag{39}$$

The solution of (39), under the hypothesis of Theorem 1.4.3, is of the form

$$x=G*f, \tag{40}$$

with $G=1/P(S)\in\mathfrak{X}_{D^{n-1}}$, which is the transfer function or the algebraic Green function of the problem (39).

1.4.5. Solvability of a Taylor boundary value problem in a resonance case. We shall study the general boundary value problem (30), (31) with Taylor boundary value conditions in the so-called resonance case, i. e. when some of the zeros of the polynomial $P(\lambda)$ are eigenvalues of the elementary boundary value problem $Dy-\lambda y=0$, $Fy=0$. Here F denotes the corresponding defining projector of a right-inverse operator L of D. Let $\lambda_1, \lambda_2,\ldots, \lambda_l$ be those zeros of $P(\lambda)$ which are eigenvalues of the problem $Dy-\lambda y=0$, $Fy=0$. This is equivalent to requirement their reciprocals $1/\lambda_1, 1/\lambda_2,\ldots, 1/\lambda_l$ to be eigenvalues of the operator L. We suppose that each of the numbers $1/\lambda_k$ is a simple eigenvalue of L, i. e. that the subspace $\{y: (1-\lambda_k L)y=0\}$ is one-dimensional for $k=1, 2, \ldots, l$ and that the corresponding spectral multiplicities $\mu_1, \mu_2, \ldots, \mu_l$ are finite.

The main assumption on which the following considerations are based is that of the existence of an annihilators-free convolution $*$ of L in $\mathfrak{X}$. Theorem 1.3.5 ensures the existence of elements $u_k \in \mathfrak{X}_{\lambda_k;\, \mu_k} = \ker(1-\lambda_k L)^{\mu_k}$ such that the operators $P_{\lambda_k} x = x * u_k$ are projectors of $\mathfrak{X}$ on $\mathfrak{X}_{\lambda_k;\, \mu_k}$.

In other terms, $u_k * u_k = u_k$ for $k=1, 2, \ldots$. The elements

$$u_k^{(1)} = (1-\lambda_k L)^{\mu_k - 1} u_k, \quad u_k^{(2)} = (1-\lambda_k L)^{\mu_k - 2} u_k, \ldots, \quad u_k^{(\mu_k)} = u_k$$

form a basis in $\mathfrak{X}_{\lambda_k;\, \mu_k}$, and $u_k^{(1)}$ is an eigenelement of L with the eigenvalue $1/\lambda_k$, i. e. $(1-\lambda_k L)u_k^{(1)} = 0$.

Definition 5. *Residual subspace*, corresponding to the eigenvalues $\lambda_1, \lambda_2, \ldots, \lambda_l$ of the eigenvalue problem $Dy-\lambda y=0$, $Fy=0$, is said to be the subspace

(41) $$\tilde{\mathfrak{X}}_{\lambda_1, \ldots, \lambda_l} = \{x \in \mathfrak{X}: x * u_k = 0, \quad k=1, 2, \ldots, l\},$$

where $u_1, u_2, \ldots, u_l$ are the elements determining the multiplier projectors $P_{\lambda_k} x = x * u_k$ on the eigensubspaces $\mathfrak{X}_{\lambda_k;\, \mu_k}$ of L.

Lemma 2. *If $\lambda_1, \lambda_2, \ldots, \lambda_l$ are different simple eigenvalues of the problem $Dy-\lambda y=0$, $Fy=0$, with finite spectral multiplicities μ_k, then the decomposition*

(42) $$\mathfrak{X} = \mathfrak{X}_{\lambda_1;\, \mu_1} \oplus \mathfrak{X}_{\lambda_2;\, \mu_2} \oplus \ldots \oplus \mathfrak{X}_{\lambda_l;\, \mu_l} \oplus \tilde{\mathfrak{X}}_{\lambda_1, \ldots, \lambda_l}$$

holds.

Proof. If we denote

$$\tilde{x} = x - x * u_1 - x * u_2 - \ldots - x * u_l,$$

then according to Theorem 1.3.4, $\tilde{x} * u_k = 0$ for $k = 1, 2, \ldots, l$, i. e. $\tilde{x} \in \tilde{\mathfrak{X}}_{\lambda_1, \ldots, \lambda_l}$. Hence each element $x \in \mathfrak{X}$ can be represented in the form

(43) $$x = x_1 + x_2 + \ldots + x_l + \tilde{x},$$

where $x_k \in \mathfrak{X}_{\lambda_k;\, \mu_k}$, $k=1, 2, \ldots, l$, and $\tilde{x} \in \tilde{\mathfrak{X}}_{\lambda_1, \ldots, \lambda_l}$.

It remains to prove that the representation (43) is unique. Indeed, if (43) is a representation of x of the form considered, then $x * u_k = x_k * u_k = x_k$, since the operator $P_{\lambda_k} x = x * u_k$ is a projector of $\mathfrak{X}$ on $\mathfrak{X}_{\lambda_k;\, \mu_k}$. Therefore, $x_1, \ldots, x_l$ and $\tilde{x}$ are uniquely determined too. The lemma is proved. □

Now let us consider the Taylor boundary value problem

(44) $$P(D)y=f, \quad FD^k y = \gamma_k \in \ker D, \quad k=0, 1, \ldots, n-1,$$

under the assumption that l different zeros $\lambda_1, \lambda_2, \ldots, \lambda_l$ of the polynomial $P(\lambda)$ are simple eigenvalues of the problem $Dy-\lambda y=0$, $Fy=0$, with finite spectral multiplicities $\mu_1, \mu_2, \ldots, \mu_l$. For simplicity sake, we shall confine our consideration to homogeneous boundary value conditions only, i. e. with the problem

(45) $$P(D)y=f, \quad FD^k y=0, \quad k=0, 1, \ldots, n-1.$$

This does not cause any loss of generality, since each boundary value problem (44) can be reduced by a simple change of the indeterminate to a problem of the form (45), but with another right-hand side f. Indeed, from the general Taylor formula (33) we have

$$y=Fy+LFDy+\dots+L^{n-1}FD^{n-1}y+L^nD^ny.$$

By means of the substitution

(46) $$y=\gamma_0+L\gamma_1+\dots+L^{n-1}\gamma_{n-1}+u$$

the problem (44) is easily reduced to the problem

$$P(D)u=f-\sum_{k=0}^{n-1}P(D)L^k\gamma_k,\quad FD^ku=0,\quad k=0,1,\dots,n-1.$$

That is why we restrict ourselves to problem (45) only. As we have already seen, in this case the problem can be reduced to a single equation $P(S)y=f$ in the multiplier quotients ring M of $(\mathfrak{X}, *)$.

Theorem 1.4.4. *Problem* (45) *has a solution in* $\mathfrak{X}$ *iff each of the equations*

(47) $$P(D)y_k=f*u_k,\quad k=1,2,\dots,l$$

has a solution in $\mathfrak{X}_{\lambda_k;\mu_k}$, *and the problem*

(48) $$P(D)\tilde{y}=\tilde{f},\quad FD^k\tilde{y}=0,\quad k=0,1,\dots,n-1,$$

with $\tilde{f}=f-f*u_1-\dots-f*u_l$, *has a solution in* $\tilde{\mathfrak{X}}_{\lambda_1,\lambda_2,\dots,\lambda_l}$.

Proof. Obviously, the elements of the eigenspaces $\mathfrak{X}_{\lambda_k;\mu_k}$, $k=1,2,\dots,l$, of L satisfy the boundary value condition $FD^my=0$ for every integer $m\geq 0$. All these subspaces are invariant for D, and hence, for $P(D)$ too. If boundary value problem (45) has a solution y, then, by Lemma 3, the representation

$$y=y_1+y_2+\dots+y_l+\tilde{y},$$

with $y_k=y*u_k$, $k=1,2,\dots,l$, is a direct sum. Therefore

$$P(D)y=P(D)y_1\oplus\dots\oplus P(D)y_l\oplus P(D)\tilde{y}$$

and hence y_k, $k=1,2,\dots,l$, are solutions of (47) in $\mathfrak{X}_{\lambda_k;\mu_k}$, and $\tilde{y}$ is a solution of $P(D)\tilde{y}=\tilde{f}$ in $\tilde{\mathfrak{X}}_{\lambda_1,\lambda_2,\dots,\lambda_l}$, such that $FD^k\tilde{y}=0$, $k=0,1,\dots,n-1$.

Conversely, if $y_1, y_2,\dots,y_l$ are solutions of (47) from the corresponding eigenspaces $\mathfrak{X}_{\lambda_k;\mu_k}$, and $\tilde{y}$ is a solution of boundary value problem (47) from $\tilde{\mathfrak{X}}_{\lambda_1,\lambda_2,\dots,\lambda_l}$, then $y=y_1+\dots+y_l+\tilde{y}$ gives a solution of the problem (45). □

The assumption that the eigenspaces $\mathfrak{X}_{\lambda_k;\mu_k}$ of L are finite-dimensional (μ_k-dimensional) implies that each of the equations $P(D)y_k=f*u_k$, $k=1,2,\dots,l$ is equivalent to a system of linear simultaneous equations. One can at once point out necessary and sufficient conditions for solvability of these systems. For the solution of these systems linear algebra methods should be used. In the simplest case, the corresponding eigenspaces are one-dimensional. Let $\mathfrak{X}_{\lambda_k;\mu_k}$ be one-dimensional, i. e. $\mu_k=1$, and let u_k be an eigenelement of L with the eigenvalue $1/\lambda_k$. Then $f*u_k$ is an eigenelement of L with the same eigenvalue $1/\lambda_k$ too. Since the corresponding eigenspace is one-dimensional, then

$$f*u_k=\chi_k(f)u_k. \tag{49}$$

The coefficient $\chi_k(f)$ is a linear functional, which is appropriate to be called the Fourier coefficient of f with respect to the eigenelement u_k.

Obviously, a necessary and sufficient condition for the solvability of the equation $P(D)y_k=\chi_k(f)u_k$ in $\mathfrak{X}_{\lambda_k;1}$ is $\chi_k(f)=0$. Then all solutions of this equation are contained in the formula $y_k=Cu_k$, where C is an arbitrary constant.

Now we shall show that the Taylor boundary value problem (48) in $\tilde{\mathfrak{X}}_{\lambda_1,\lambda_2,\dots,\lambda_l}$ is not different in essence from the non-resonance Taylor boundary value problem (30), (31) already considered.

The subspace $\tilde{\mathfrak{X}}_{\lambda_1,\lambda_2,\dots\lambda_l}$ with the multiplication operation $*$ is a subalgebra of the convolutional algebra $(\mathfrak{X},*)$. The restriction of the right inverse operator $Lx=r*x$ of D to $\tilde{\mathfrak{X}}_{\lambda_1,\lambda_2,\dots,\lambda_r}$ is a convolutional operator in $\mathfrak{X}_{\lambda_1,\lambda_2,\dots,\lambda_l}$ too, and it can be represented in the form $\tilde{L}\tilde{x}=\tilde{r}*\tilde{x}$ with

$$\tilde{r}=r-r*u_1-r*u_2-r*u_3-\dots-r*u_l.$$

Since none of the zeros of the polynomial $P(\lambda)$ is any longer an eigenvalue of the problem $D\tilde{y}-\lambda\tilde{y}=0, F\tilde{y}=0$, in $\mathfrak{X}_{\lambda_1,\dots,\lambda_l}$, a sufficient condition for the solvability of the boundary value problem (48) in $\mathfrak{X}_{\lambda_1,\lambda_2,\dots,\lambda_l}$ is given by Theorem 1.4.3.

Let $\tilde{\mathsf{M}}=\tilde{\mathsf{M}}_{\lambda_1,\lambda_2,\dots,\lambda_l}$ be the convolution quotients ring of $(\tilde{\mathfrak{X}}_{\lambda_1,\lambda_2,\dots,\lambda_l},*)$, and let $\tilde{S}=1/\tilde{L}\in\tilde{\mathsf{M}}$.

Then, according to Theorem 1.4.3, for the solvability of the boundary value problem (48) in $\tilde{\mathfrak{X}}_{\lambda_1,\lambda_2,\dots,\lambda_l}$ it is sufficient each of the partial fractions $1/(S-\lambda_k)$, where λ_k is a zero of the polynomial $P(\lambda)$, to be an element of $\tilde{\mathfrak{X}}_{\lambda_1,\lambda_2,\dots,\lambda_l}$. Then the solution sought $\tilde{y}$ has the form

$$\tilde{y}=\tilde{G}*\tilde{f} \tag{50}$$

with $\tilde{G}=1/P(S)$.

The general scheme for treating Taylor boundary value problems for abstract differential equation in the resonance case, considered here, is due in essence to Grozdev [63]. Similar considerations, but in the more special situation of a Hilbert space $\mathfrak{X}$, of a complete system of eigenelements of L in $\mathfrak{X}$ with one-dimensional eigenspaces, are made by Tasche [96].

CHAPTER 2

CONVOLUTIONS OF GENERAL INTEGRATION OPERATORS. APPLICATIONS

In this chapter we consider systematically the problem of finding convolutions for an arbitrary right inverse operator of the differentiation operator in various function spaces, in particular for continuous, for locally integrable and for locally holomorphic functions. The resulting convolutions are used for an algebraic approach to the problem of developing locally holomorphic functions in Dirichlet series on a given exponent system. Thus we get the Leontiev expansion. A new problem solved by this algebraic approach is the multiplier problem for the formal Leontiev expansions of locally holomorphic functions. The convolutional approach is extended to some analogons of the general integration operators: the right inverses of the backward shift operator and Gelfond-Leontiev integration operator.

2.1. CONVOLUTIONS OF THE LINEAR RIGHT INVERSES OF THE DIFFERENTIATION OPERATOR

The Volterra integration operator $lf(t)=\int_0^t f(\tau)d\tau$ is only a very special kind of a right inverse operator of the differentiation operator. In an example considered in 1.3.4 we have shown that an arbitrary right inverse operator of $D=d/dt$ in $\mathscr{C}([0,1])$ has the form $Lf(t)=lf(t)+\Phi(f)$ with an arbitrary linear functional Φ on $\mathscr{C}([0,1])$. It is quite natural to ask for a convolution of L in $\mathscr{C}([0,1])$.

Till quite recently, no convolutions for such a general right inverse operator of L were known. Here we shall present in an explicit form several convolutions of such a right inverse operator.

2.1.1. A class of convolutions, depending on an arbitrary linear functional in spaces of continuous functions. Let Δ be an arbitrary interval (finite, infinite, closed, open, or half-open) and let $\mathscr{C}(\Delta)$ be the space of the complex-valued continuous functions on Δ.

We consider an arbitrary, but fixed, linear functional Φ on $\mathscr{C}(\Delta)$, continuous in the topology of $\mathscr{C}(\Delta)$, i. e. with respect to the almost uniform convergence in $\mathscr{C}(\Delta)$. According to the Riesz-Markov theorem (Edwards [55], sect. 4.10.1), each linear functional in $\mathscr{C}(\Delta)$ has a compact support, i. e. there exists a finite segment $[\alpha, \beta] \subset \Delta$ and functions $\mu_j(t)$, $j=1, 2$, with bounded variation in $[\alpha, \beta]$, such that the functional Φ has the representation

$$(1) \qquad \Phi(f)=\int_{\alpha}^{\beta} f(x)d\mu(x)$$

with $\mu(x)=\mu_1(x)+i\mu_2(x)$, and the integral is to be understood in Riemann-Stieltjes sense. A shorter and universal way of stating the *Riesz-Markov theorem* is the following: For each continuous linear functional $\Phi(f)$ on $\mathscr{C}(\Delta)$ there exists a unique complex Radon measure μ on Δ with a compact support on Δ, such that

$$(1') \qquad \Phi(f)=\int_{\Delta} f(t)d\mu(t).$$

Conversely, if μ is a complex Radon measure on Δ with compact support then (1′) defines a continuous linear functional on $\mathscr{C}(\Delta)$.

In this formulation, the theorem is valid not only for an interval but for an arbitrary locally compact space. However, for our purposes the above statement of the Riesz-Markov theorem in classical terms is enough. The uniqueness of representation (1) can be ensured by the assumption that $\mu(t)$ is continuous from right or left, and by fixing a value of μ in a point of Δ.

Theorem 2.1.1. *If Φ: $\mathscr{C}(\Delta) \to \mathbf{C}$ is an arbitrary non-zero continuous linear functional on $\mathscr{C}(\Delta)$, then the operation*

$$(2) \qquad (f*g)(t)=\Phi_x\left\{\int_{x}^{t} f(x+t-\tau)g(\tau)d\tau\right\},$$

where the subscript x on Φ means that the functional Φ acts on the variable x only, is a bilinear, commutative and associative operation in $\mathscr{C}(\Delta)$. It is separately continuous in $\mathscr{C}(\Delta)$ and the algebra $(\mathscr{C}(\Delta), *)$ is annihilators-free.

Proof. It is evident that the operation $*$ is an inner operation in $\mathscr{C}(\Delta)$, i. e. $*$: $\mathscr{C}(\Delta)\times\mathscr{C}(\Delta)\to\mathscr{C}(\Delta)$. The continuity of (2) follows easily from the almost evident inequality

$$(3) \qquad \|f*g\|_K \leqq A_K . |\mu|(K) . \|f\|_K . \|g\|_K,$$

where $K=[a, b]$ is a compact subinterval of Δ, $\|h\|_K=\sup\{|h(t)|, t\in K\}$, and $|\mu|(K)=\operatorname{Var}\mu_1+\operatorname{Var}\mu_2$ with $A_K=\max(b, \beta)-\min(a, \alpha)$. From (3) it follows not only the separate continuity of $f*g$ but continuity in the sense that $f_n\to f$ and $g_n\to g$ imply $f_n*g_n\to f*g$.

The commutativity and bilinearity of (2) are evident. Let us prove the associativity. For the sake of brevity, let us introduce the entire function of exponential type

$$(4) \qquad E(\lambda)=\Phi_x(e^{\lambda x}),$$

which we shall name the *indicatrix* of the functional Φ. For arbitrary $\lambda, \mu \in \mathbf{C}$, $\lambda \neq \mu$ we have

$$\{e^{\lambda t}\} * \{e^{\mu t}\} = \left\{\frac{E(\lambda)e^{\mu t} - E(\mu)e^{\lambda t}}{\mu - \lambda}\right\}, \tag{5}$$

as it can be seen by direct computation. If $\nu \in \mathbf{C}$, $\nu \neq \lambda$ and $\nu \neq \mu$, but otherwise arbitrary, then (5) gives

$$(\{e^{\lambda t}\} * \{e^{\mu t}\}) * \{e^{\nu t}\} = \frac{E(\mu)E(\nu)}{(\mu-\lambda)(\nu-\lambda)} e^{\lambda t} + \frac{E(\nu)E(\lambda)}{(\nu-\mu)(\lambda-\mu)} e^{\mu t} + \frac{E(\lambda)E(\mu)}{(\lambda-\nu)(\mu-\nu)} e^{\nu t}. \tag{6}$$

This expression is invariant for cyclic permutations of λ, μ, ν and therefore

$$(\{e^{\lambda t}\} * \{e^{\mu t}\}) * \{e^{\nu t}\} = \{e^{\lambda t}\} * (\{e^{\mu t}\} * \{e^{\nu t}\}). \tag{7}$$

Let us now differentiate (7) l times with respect to λ, m times with respect to μ and n times with respect to ν. From expression (2) it is clear that the differentiation under the functional is allowed, and hence

$$(\{t^l e^{\lambda t}\} * \{t^m e^{\mu t}\}) * \{t^n e^{\nu t}\} = \{t^l e^{\lambda t}\} * (\{t^m e^{\mu t}\} * \{t^n e^{\nu t}\})$$

for all $l, m, n = 0, 1, 2, \ldots$. If $\lambda \to 0$, $\mu \to 0$, and $\nu \to 0$, in a way that $\lambda \neq \mu \neq \nu \neq \lambda$, then we get

$$(\{t^l\} * \{t^m\}) * \{t^n\} = \{t^l\} * (\{t^m\} * \{t^n\}) \tag{8}$$

for $l, m, n = 0, 1, 2, \ldots$. From the bilinearity of the operation $*$ it follows that the associativity relation

$$(f * g) * h = f * (g * h) \tag{9}$$

holds for polynomials f, g and h. From the continuity of (2) in $\mathscr{C}(\Delta)$ and from the denseness of the polynomials in $\mathscr{C}(\Delta)$ it follows that (9) is valid for arbitrary f, g and $h \in \mathscr{C}(\Delta)$. Hence (2) is an associative operation.

Finally, let us prove that the algebra $(\mathscr{C}(\Delta), *)$ is annihilators-free. We can show even something more: that there is a non-divisor of zero of (2), different from 0. Indeed, let $\lambda_0 \in \mathbf{C}$ be chosen so that $\Phi_x(e^{\lambda_0 x}) = E(\lambda_0) \neq 0$. From the hypothesis that Φ is a non-zero functional it follows that such a λ_0 exists, since otherwise we should have $\Phi(e^{\lambda x}) = 0$ for every $\lambda \in \mathbf{C}$, and hence $\Phi_x(x^l e^{\lambda x}) = 0$ for all $l = 0, 1, 2, \ldots$. Then if we let $\lambda \to 0$, we would have $\Phi(x^l) = 0$ and therefore $\Phi(f) = 0$ for polynomials f. If $f \in \mathscr{C}(\Delta)$ is arbitrary, then taking a sequence of polynomials f_n, converging to f in the topology of $\mathscr{C}(\Delta)$, we would get $\Phi(f) = 0$, in contradiction to the assumption that Φ is a non-zero functional.

Let $\lambda_0 \in \mathbf{C}$ be such that $E(\lambda_0) \neq 0$ and let us consider the convolutional operator

$$L_{\lambda_0} f(t) = \left\{\frac{e^{\lambda_0 t}}{E(\lambda_0)}\right\} * f(t).$$

Using expression (2) for the convolution $*$, we represent L_{λ_0} in the form

$$L_{\lambda_0} f(t) = e^{\lambda_0 t} \int_a^t e^{-\lambda_0 \tau} f(\tau)\, d\tau - \frac{e^{\lambda_0 t}}{E(\lambda_0)} \Phi_x\left\{e^{\lambda_0 x} \int_a^x e^{-\lambda_0 \tau} f(\tau)\, d\tau\right\}. \tag{10}$$

It is easy to see that L_{λ_0} is a right inverse operator of the differential operator $D_{\lambda_0}=(d/dt)-\lambda_0$, determined by the boundary value condition $\Phi(L_{\lambda_0}f)=0$. Then it is not difficult to see that the function $e^{\lambda_0 t}$ is not a divisor of zero of the convolution $*$. In fact, if $\{e^{\lambda_0 t}\}*f=0$ for a $f\in\mathscr{C}(\Delta)$, then this is equivalent to the equation $L_{\lambda_0}(E(\lambda_0)f)=0$. Then, if we apply to the last equation the operator $D_{\lambda_0}=(d/dt)-\lambda_0$, we get $E(\lambda_0)f(t)=0$, or $f(t)\equiv 0$.

Thus the theorem is proved. □

Corollary. If $\Phi: \mathscr{C}^1(\Delta)\to\mathbf{C}$ is an arbitrary linear non-zero functional on the space $\mathscr{C}^1(\Delta)$ of the smooth functions in Δ, then operation (2) converts $\mathscr{C}^1(\Delta)$ into an annihilators-free commutative and associative algebra with separately continuous multiplication (2) in $\mathscr{C}^1(\Delta)$.

Proof. It is easy to see that an arbitrary linear functional $\Phi(f)$ on $\mathscr{C}^1(\Delta)$ has the form

$$\Phi(f)=\Psi(f')+\Phi(1)f(0) \tag{11}$$

where Ψ is a linear functional on $\mathscr{C}(\Delta)$. Then

$$(f*g)'(t)=\Phi(f)g(t)+\Phi_x\left\{\int_x^t f'(x+t-\tau)g(\tau)d\tau\right\},$$

and hence $(f*g)'\in\mathscr{C}(\Delta)$, thus showing that (2) is an inner operation in $\mathscr{C}^1(\Delta)$. The associativity of (2) in $\mathscr{C}(\Delta)$ can be seen in the same way as in the proof of the previous theorem. The same is the proof of the assertion that the algebra $[\mathscr{C}^1(\Delta), *]$ is annihilators-free.

Now, we give some useful properties of operation (2).

Lemma 1. *If $*$ is operation (2), then*

$$\Phi(f*g)=0 \quad \textit{for all} \quad f, g\in\mathscr{C}(\Delta). \tag{12}$$

Proof. Denoting

$$h(x, t)=\int_x^t f(x+t-\tau)g(\tau)d\tau, \tag{13}$$

we have $h(x, t)=-h(t, x)$. Then $\Phi(f*g)=\Phi_t\Phi_x\{h(x, t)\}=-\Phi_t\Phi_x\{h(t, x)\}$ $=-\Phi_x\Phi_t\{h(t, x)\}=-\Phi(f*g)$ and hence $\Phi(f*g)=0$. We have used only the Fubini property $\Phi_t\Phi_x=\Phi_x\Phi_t$. □

Lemma 2. *If $f, g\in\mathscr{C}(\Delta)$, and if g is a function with (locally) bounded variation, then $f*g\in\mathscr{C}^1(\Delta)$.*

Proof. For a fixed $x\in\Delta$ the operation

$$(f\overset{x}{*}g)(t)=\int_x^t f(t+x-\tau)g(\tau)d\tau$$

is again the Duhamel convolution of f and g, but with the initial point $t=x$, instead of $t=0$. In fact, it can be obtained from the Duhamel convolution

$$f\tilde{*}g(t)=\int_0^t f(t-\tau)g(\tau)d\tau$$

by means of the transformation T: $f(t)\to f(x+t)$, using the formula $f\overset{(x)}{*}g=T^{-1}[(Tf)\tilde{*}(Tg)]$. Then, according to Lemma 5 of sect. 1.1, the function $h(x, t)$ with fixed x belongs to $\mathscr{C}^1(\Delta)$, and

$$\frac{\partial h}{\partial t}=\int_x^t f(x+t-\tau)dg(\tau)+f(t)g(x).$$

The last identity can be shown easily directly too. Then, integrating it from a to t, we get

$$h(x,t)=h(x,a)+\int_a^t du\int_x^u f(u+x-\tau)dg(\tau)+g(x)\int_a^t f(u)du.$$

Thus

$$(f*g)(t)=\Phi_x\{h(x,a)\}+\int_a^t \Phi_x\left\{\int_x^u f(u+x-\tau)dg(\tau)+g(x)f(u)\right\}du$$

and hence $f*g\in\mathscr{C}^1(\Delta)$. For the derivative $(f*g)'$ we obtain the expression

$$(f*g)'(t)=\Phi_x\left\{\int_x^t f(x+t-\tau)dg(\tau)\right\}+\Phi(g)f(t). \tag{14}$$

Thus the lemma is proved. □

Now we show that operation (2) can be extended to an operation in the space of locally integrable functions $\mathscr{L}(\Delta)$ on Δ as a continuous, bilinear, commutative and associative operation. Moreover, this can be done in only one way. To this end we shall prove two auxiliar propositions.

Lemma 3. *Let $f\in\mathscr{L}(\Delta)$ and g be absolutely continuous (at least locally) on Δ. Then for a fixed $x\in\Delta$ the function*

$$h(x,t)=\int_x^t f(t+x-\tau)g(\tau)d\tau$$

is an absolutely continuous function of t in Δ.

Let us note that under "absolutely continuous" we mean "absolutely continuous on every compact subinterval of Δ". There is no need to give a proof here, since an analogous proposition is proved by Mikusiński and Ryll-Nardzewski [80], Theorem 6, since for a fixed x the expression for $h(x, t)$ is only a modification of the Duhamel convolution. An equivalent statement of the lemma is to say that $h(x, t)$ as a function of t in Δ is locally absolutely continuous.

Lemma 4. *Let $f\in\mathscr{L}(\Delta)$, and g be locally absolutely continuous in Δ. Then the function $f*g$, defined by* (2), *is locally absolutely continuous and the differentiation formula*

$$(f*g)'(t)=\Phi_x\left\{\int_x^t f(t+x-\tau)g'(\tau)d\tau\right\}+\Phi(g)f(t). \tag{15}$$

holds.

Proof. From the assumption that $g(t)$ is locally absolutely continuous in Δ it follows that $g(t)=g(x)+\int_x^t g'(u)du$. Then, substituting this representation of $g(t)$ in $h(x, t)$, we get

$$h(x, t)=g(x)\int_x^t f(u)du+\int_x^t\left[\int_x^u f(u+x-\tau)g'(u)du\right]d\tau.$$

Here we have used the fact that $h(x, t)$ for fixed $x\in\Delta$, is a convolution of the operator $l_x f(t)=\int_x^t f(u)du$. Representing the integral $\int_x^t$ as $\int_x^a+\int_a^t$, we get

$$(16)\qquad (f*g)(t)=\text{const}+\Phi_x\left\{\int_a^t\left(f(u)g(x)+\int_x^u f(u+x-\tau)g'(\tau)d\tau\right)du\right\}$$

and the assertion of the lemma follows immediately. □

Theorem 2.1.2. *The operation, defined by* (2) *in* $\mathscr{C}(\Delta)$, *admits a unique continuous extension to the space* $\mathscr{L}(\Delta)$ *of the locally integrable functions in* Δ.

Proof. Let the number $\lambda_0\in\mathbf{C}$ be chosen so that $E(\lambda_0)=\Phi_x\{e^{\lambda_0 x}\}\neq 0$. In the proof of Theorem 2.1.1, we have seen that this is always possible, provided Φ is non-zero linear functional. Here we also introduce the auxiliar convolutional operator

$$(17)\quad L_{\lambda_0}f(t)=\left\{\frac{e^{\lambda_0 t}}{E(\lambda_0)}\right\}*f(t)=e^{\lambda_0 t}\int_a^t e^{-\lambda_0\tau}f(\tau)d\tau-\frac{e^{\lambda_0 t}}{E(\lambda_0)}\Phi_x\left\{e^{\lambda_0 x}\int_a^x e^{-\lambda_0\tau}f(\tau)d\tau\right\}.$$

It is a right inverse operator of the differential operator $D_{\lambda_0}=(d/dt)-\lambda_0$, with the property $\Phi(L_{\lambda_0}f)=0$. If $f\in\mathscr{C}(\Delta)$, then $L_{\lambda_0}f$, defined by (17), is a locally absolutely continuous function in Δ. If $f, g\in\mathscr{C}(\Delta)$, then

$$(18)\qquad f*g=D_{\lambda_0}[(L_{\lambda_0}f)*g].$$

But according to Lemma 4, the right-hand side of (18) is defined for arbitrary $f, g\in\mathscr{L}(\Delta)$ too. Therefore, we can define the extension of the operation $*$ to $\mathscr{L}(\Delta)$ as

$$(18')\qquad f\,\tilde{*}\,g\overset{\text{def}}{=}D_{\lambda_0}[(L_{\lambda_0}f)*g].$$

Let us show that the operation $\tilde{*}$ defined by (18′) in $\mathscr{L}(\Delta)$ is a separately continuous operation. Indeed, let $f_n\to f$ and $f_n\,\tilde{*}\,g\to h$ in $\mathscr{L}(\Delta)$. Then $L^2_{\lambda_0}(f_n\,\tilde{*}\,g)=L_{\lambda_0}f_n*L_{\lambda_0}g\to L^2_{\lambda_0}h$ not only in $\mathscr{L}(\Delta)$ but in $\mathscr{C}(\Delta)$ too. Since $L_{\lambda_0}f_n\to L_{\lambda_0}f$ in $\mathscr{C}(\Delta)$, then from the separate continuity of $*$ in $\mathscr{C}(\Delta)$ it follows that $L^2_{\lambda_0}h=(L_{\lambda_0}f)*(L_{\lambda_0}g)$. Hence $L^2_{\lambda_0}h=L^2_{\lambda_0}(f\,\tilde{*}\,g)$, i. e. $L^2_{\lambda_0}(h-f\,\tilde{*}\,g)=0$. Therefore, $h-f\,\tilde{*}\,g=0$. According to the *closed graph theorem*, the convolutional operator $M_g f = f * g$ is continuous in $\mathscr{L}(\Delta)$. Hence operation (18′) is separately continuous. The commutativity and the associativity of $\tilde{*}$ in $\mathscr{L}(\Delta)$ follows from that of $*$ in $\mathscr{C}(\Delta)$ due to the continuity of $f \tilde{*} g$. □

Bozhinov [11] had proposed another construction of the extension of $*$ from $\mathscr{C}(\Delta)$ to $\mathscr{L}(\Delta)$ by means of the representation $f \tilde{*} g = \tilde{\Phi}\{h(x, t)\}$, where $\tilde{\Phi}$ is the *Lebesgue extension* of Φ to $\mathscr{L}(\Delta)$. Using this extension, Lemma 4 can be stated in the following way:

Lemma 4′. *If $f \in \mathscr{L}(\Delta)$ and g is locally absolutely continuous in Δ, then the function $f \tilde{*} g = \tilde{\Phi}_x \{h(x, t)\}$, where $\tilde{\Phi}$ is the Lebesgue extension of Φ, is locally absolutely continuous and the differentiation formula*

$$(19) \qquad (f \tilde{*} g)'(t) = \tilde{\Phi}_x \left\{ \int_x^t f(t + x - \tau) g'(\tau) d\tau \right\} + \Phi(g) f(t)$$

holds.

To Bozhinov [11] is due a stronger result:

If $f, g \in \mathscr{L}(\Delta)$ and g is almost everywhere equal to a function with locally bounded variation in Δ, then $f \tilde{*} g$ is a locally absolutely continuous function.

An analogous assertion for the Duhamel convolution had been proved in sect. 1.1. Here we shall not give a proof of this more general proposition.

Let us return to operation (2) in the space $\mathscr{C}(\Delta)$. We shall prove a property of $*$ which we shall use later.

Lemma 5. *Let the linear functional $\Phi: \mathscr{C}(\Delta) \to \mathrm{C}$ admit representation in the form*

$$(20) \qquad \Phi(f) = \Psi_x \left\{ \int_a^x f(\tau) d\tau \right\}$$

with a linear functional $\Psi: \mathscr{C}(\Delta) \to \mathrm{C}$ and $a \in \Delta$. Then the operation $*$, defined by (2), maps $\mathscr{C}(\Delta) \times \mathscr{C}(\Delta)$ into $\mathscr{C}^1(\Delta)$ and

$$(21) \qquad (f * g)'(t) = f(t)\Phi(g) + g(t)\Phi(f)$$

$$- \Psi(1) \int_a^t f(a + t - \tau) g(\tau) d\tau + \Psi_x \left\{ \int_x^t f(t + x - \tau) g(\tau) d\tau \right\}.$$

Proof. Since $\int_a^x = \int_a^t - \int_x^t$, we can transform $\int_a^x h(u, t) du$ in the following way. We have

$$\int_a^x h(\xi, t) d\xi = k(x, t) + l(x, t)$$

with

$$k(x,t)=\int_a^t du\int_u^t f(t+u-\tau)g(\tau)d\tau$$

$$=\int_a^t g(\tau)d\tau\int_a^\tau f(t+u-\tau)du=\int_a^t g(\tau)d\tau\int_{t+a-\tau}^t f(\sigma)d\sigma$$

and

$$l(x,t)=\int_t^x du\int_u^t f(t+u-\tau)g(\tau)d\tau=-\int_t^x\left[\int_t^{t+x-\tau} f(\sigma)d\sigma\right]g(\tau)d\tau.$$

These expressions for $k(x,t)$ and $l(x,t)$ exhibit that they are differentiable with respect to t. We get

$$\frac{\partial k}{\partial t}=g(t)\int_a^t f(\sigma)d\sigma+f(t)\int_a^t g(\sigma)d\sigma-\int_a^t f(t+a-\tau)g(\tau)d\tau$$

and

$$\frac{\partial l}{\partial t}=g(t)\int_t^x f(\sigma)d\sigma+f(t)\int_t^x g(\sigma)d\sigma+\int_x^t f(t+x-\tau)g(\tau)d\tau.$$

Hence the function $(f*g)(t)=\Psi_x\left\{\int_a^x h(u,t)du\right\}$ is differentiable and

$$(f*g)'(t)=f(t)\Phi(g)+g(t)\Phi(f)-\Psi(1)\int_a^t f(t+a-\tau)g(\tau)d\tau$$

$$+\Psi_x\left\{\int_x^t f(t+x-\tau)g(\tau)d\tau\right\}.$$

The lemma is proved. □

2.1.2. Linear right inverses of the differentiation and their convolutions. Here we consider the most general linear and continuous right inverse operator of the differentiation operator $D=d/dt$ in the spaces $\mathscr{C}(\Delta)$ and $\mathscr{L}(\Delta)$ of the continuous and of the locally integrable functions in an interval Δ, and in the space $\mathscr{H}(\overline{D})$ of locally holomorphic functions in a finite, closed and convex domain $\overline{D}$ of the complex plane. Since the considerations in $\mathscr{C}(\Delta)$ and in $\mathscr{L}(\Delta)$ proceed in a similar way, we restrict them to the space $\mathscr{C}(\Delta)$. As for the space $\mathscr{H}(\overline{D})$, here we encounter some peculiarities which justify its separate consideration.

For the sake of brevity we shall use the term "integration operator" instead of the term "right inverse operator of the differentiation operator".

From the context it should be clear which right inverse of the differentiation is meant.

Lemma 6. *A linear operator* $L: \mathscr{C}(\Delta) \to \mathscr{C}(\Delta)$ *is an integration operator iff it admits a representation of the form*

$$Lf(t) = \int_a^t f(\tau)d\tau + \Psi(f), \tag{22}$$

where a is an arbitrary point of the interval Δ, *and* $\Psi: \mathscr{C}(\Delta) \to \mathbf{C}$ *is a linear functional.*

Proof. It is clear that each operator of the form (22) is a linear right inverse of the differentiation operator. Conversely, let $L: \mathscr{C}(\Delta) \to \mathscr{C}(\Delta)$ be a linear right inverse of $D = d/dt$, i. e. $L: \mathscr{C}(\Delta) \to \mathscr{C}^1(\Delta)$ and $(d/dt)Lf = f$ for every $f \in \mathscr{C}(\Delta)$. The last condition means that $Lf(t)$ is a primitive of the function $f(t)$. But each primitive of $f(t)$ has the form

$$\int_a^t f(\tau)d\tau + A_f, \tag{23}$$

where A_f is a constant, possibly dependent on f. If (23) is the representation of $Lf(t)$, from the linearity of Lf it follows that $A_{f+g} = A_f + A_g$ and $A_{cf} = cA_f$, i. e. that A_f is a linear functional in $\mathscr{C}(\Delta)$. The lemma is proved. □

In 1.3.3 we had given some elements of the theory of the linear right inverse operators. The first and most essential notion connected with a pair D, L of a linear operator and its right inverse is that of the defining projector of L. There we had shown that the restriction of this projector to the domain of D is the operator $I - LD$. In our case of operator (22) for this restriction of the defining projector F we get

$$Ff(t) = f(t) - Lf'(t) = f(0) - \Psi(f'). \tag{24}$$

Here F is a linear functional, since in general $F: \mathscr{X}_D \to \ker D$ and in our case $\ker D$ is the space of the constant functions.

Conversely, if $F: \mathscr{C}^1(\Delta) \to \mathbf{C}$ is a projector, i. e. a linear functional with $F\{1\} = 1$, then there exists a right inverse operator L of d/dt in $\mathscr{C}(\Delta)$ such that

$$Lf(t) = \int_a^t f(\tau)d\tau - F_x\left\{\int_a^x f(\tau)d\tau\right\}. \tag{25}$$

Therefore, along with (22), each integration operator in $\mathscr{C}(\Delta)$ has a representation of the form (25) with a linear functional $F: \mathscr{C}^1(\Delta) \to \mathbf{C}$ with $F\{1\} = 1$.

Definition 1. Let $\Phi: \mathscr{C}^{(k)}(\Delta) \to \mathbf{C}$ be a linear functional in the space $\mathscr{C}(\Delta)$ of k-times differentiable functions on Δ with continuous k-th derivative. *Indicatrix* (or *exponential indicatrix*) of the linear functional Φ is said to be the entire function of the exponential type

$$E(\lambda) = \Phi_x\{e^{\lambda x}\}. \tag{26}$$

The fact that $E(\lambda)$ is really an entire function of exponential type follows from the finiteness of the support of Φ.

The *generating function* of an integration operator $L: \mathscr{C}(\Delta) \to \mathscr{C}^1(\Delta)$ with a defining projector $F: \mathscr{C}(\Delta) \to \mathbf{C}$ is named the indicatrix of the linear functional F, i. e.

$$E(\lambda) = F_x\{e^{\lambda x}\}. \tag{27}$$

Since $F\{1\} = 1$, then $E(0) = 1$.

Example 1. Since the defining projector of the Volterra integration operator $lf(t) = \int_0^t f(\tau)d\tau$ is $Ff = f(0)$, then its generating function is the constant function $E(\lambda) = 1$.

Example 2. If

$$Lf(t) = \int_0^t f(\tau)d\tau - \int_0^1 (1-\tau) f(\tau)d\tau,$$

then $Ff(t) = \int_0^1 f(\tau)d\tau$, and its generating function is

$$E(\lambda) = \int_0^1 e^{\lambda\tau}d\tau = \frac{1}{\lambda}(e^\lambda - 1).$$

One can put the converse question: If an entire function $E(\lambda)$ of exponential type with $E(0)=1$ is given, does there always exist an integration operator (25) with a generating function $E(\lambda)$?

It is easy to see that in general the answer is negative. For example, let $\Delta = [-1, 1]$, and $E(\lambda) = 1 - \lambda^2$. If there existed a linear functional F: $\mathscr{C}^1(\Delta) \to \mathbf{C}$ such that $F_x(e^{\lambda x}) = 1 - \lambda^2$, then its restriction on $\mathscr{C}^2(\Delta)$ should be unique. But the linear functional $F(f) = f(0) - f''(0)$, defined on $\mathscr{C}^2(\Delta)$, obviously has the property $F_x(e^{\lambda x}) = 1 - \lambda^2$. Then it is easy to see that there does not exist a continuous extension of the linear functional $F(f) = f(0) - f''(0)$, from $\mathscr{C}^2(\Delta)$ onto $\mathscr{C}^1(\Delta)$.

Lemma 7. *If $E(\lambda) = P_n(\lambda)e^{a\lambda}$, where $P_n(\lambda)$ is an n-th degree polynomial with $P_n(0) = 1$, and $a \in \Delta$, then $E(\lambda)$ is the indicatrix of the linear functional $F(f) = (P_n(d/dt)f)(a)$, defined on $\mathscr{C}^{(n)}(\Delta)$.*

Proof. Indeed,

$$F_x\{e^{\lambda x}\} = P_n\left(\frac{d}{dt}\right) e^{\lambda t}\Big|_{t=a} = P_n(\lambda)e^{a\lambda}.$$

In this case we can assert only that there exists a right inverse operator L of the differentiation in $\mathscr{C}^{n-1}(\Delta)$ with the generating function $E(\lambda) = P_n(\lambda)e^{a\lambda}$. It cannot be extended to a continuous operator in the whole space $\mathscr{C}(\Delta)$ if $n > 1$. □

The problem of reconstruction of the integration operator by its generating function has a complete solution in the complex domain.

Definition 2. Let D be a finite convex domain in the complex plane. Then the space $\mathscr{H}(\overline{D})$ of the locally holomorphic functions on the closure

$\bar{D}$ of D is said to be the set of the classes of analytic functions, defined in domains containing $\bar{D}$. Two functions of one class are analytical continuations of each other, coinciding on a domain containing $\bar{D}$.

The space $\mathscr{H}(\bar{D})$ is studied by S e b a s t i ã o-e- S i l v a [89], K ö t h e [69], and others. It is not a Fréchet space, but a space of the class of the DF-spaces, i. e. dual of a Fréchet space. The topology on $\mathscr{H}(\bar{D})$ is the inductive hull topology. In particular, to Sebastião-e-Silva is due the following representation theorem for the linear functionals in $\mathscr{H}(\bar{D})$:

The general form of the linear functionals on $\mathscr{H}(\bar{D})$ is

$$(28) \qquad \Phi(f) = \frac{1}{2\pi i} \int_{\Gamma_f} \gamma(z) f(z) dz,$$

where $\gamma(z)$ is a function, analytic in the complement $\bar{C} \backslash \bar{D}$ of $\bar{D}$ to the extended complex plane, and Γ_f is a contour containing $\bar{D}$ and lying in the domain of the representative f of the locally holomorphic function considered.

The representation (28) shows that the indicatrix $E(\lambda) = \Phi_z\{e^{\lambda z}\}$ of an arbitrary linear functional Φ in $\mathscr{H}(\bar{D})$ is an entire function of exponential type.

T h e o r e m 2.1.3. *If $E(\lambda)$ is an entire function of exponential type with $E(0) = 1$, then there exists a finite, closed and convex domain $\bar{D}$, such that $E(\lambda)$ is a generating function of an integration operator in $\mathscr{H}(\bar{D})$.*

P r o o f. Let

$$(29) \qquad E(\lambda) = 1 + \frac{a_1}{1!}\lambda + \frac{a_2}{2!}\lambda^2 + \cdots + \frac{a_n}{n!}\lambda^n + \cdots$$

be a given entire function of exponential type. An important notion connected with such a function is that of the Borel transform of it. The *Borel transform* of (29) (see e. g. R. B o a s. *Entire Functions*, New York, 1954, p. 73) is the function

$$(30) \qquad \gamma(z) = \frac{1}{z} + \frac{a_1}{z^2} + \frac{a_2}{z^3} + \cdots + \frac{a_n}{z^{n+1}} + \cdots$$

Let $\bar{D}$ be the smallest closed convex domain, which contains all the singularities of $\gamma(z)$. In other words, let $\bar{D}$ be the indicator diagram of $E(\lambda)$. Without any loss of generality, we can assume that $0 \in D$. It is not difficult to show that $E(\lambda)$ is the indicatrix of the linear functional

$$(31) \qquad F(f) = \frac{1}{2\pi i} \int_{\Gamma_f} \gamma(z) f(z) dz,$$

where Γ_f is a contour containing the domain $\bar{D}$ and lying in the domain of the representative f. Indeed, the identity

$$E(\lambda) = \frac{1}{2\pi i} \int_{\Gamma} \gamma(z) e^{\lambda z} dz$$

is a well-known representation of Polya (B o a s, *l. c.*, p. 74).

Then $E(\lambda)$ is the generating function of the integration operator

$$(32)\qquad Lf(z)=\int_0^z f(\zeta)d\zeta-\frac{1}{2\pi i}\int_{\Gamma_f}\gamma(\zeta)\left[\int_0^\zeta f(\sigma)d\sigma\right]d\zeta$$

in $\mathscr{H}(\bar{D})$. □

Now let us proceed to find non-trivial convolutions of the integration operators. We assume that $0\in\Delta$.

Theorem 2.1.4. *Let Φ: $\mathscr{C}(\Delta)\to\mathbf{C}$ be an arbitrary linear functional in $\mathscr{C}(\Delta)$. Then the operation*

$$(33)\qquad (f*g)(t)=\int_0^t f(t-\tau)g(\tau)d\tau+\Phi_x\left\{\frac{\partial}{\partial x}\int_t^x f(x+t-\tau)g(\tau)d\tau\right\}$$

is a convolution of the linear integration operator

$$(34)\qquad Lf(t)=\int_0^t f(\tau)d\tau+\Phi(f)$$

in $\mathscr{C}^1(\Delta)$, such that the representation

$$(35)\qquad Lf=\{1\}*f$$

holds.

Proof. We shall use the corollary from Theorem 2.1.1. The defining projector $Ff=f(0)-\Phi(f')$ of L is a linear functional in $\mathscr{C}^1(\Delta)$. Then, according to this corollary, the operation

$$(36)\qquad (f*g)(t)=F_x\left\{\int_x^t f(t+x-\tau)g(\tau)d\tau\right\}$$

is a bilinear, commutative and associative operation in $\mathscr{C}^1(\Delta)$. But (36) coincides in fact with (33). Representation (35) can be easily verified directly. Hence (33) is a convolution of L in $\mathscr{C}^1(\Delta)$. □

Now we shall show that the operator (34) has a non-trivial convolution in the whole space $\mathscr{C}(\Delta)$.

Theorem 2.1.5. *The operation*

$$(37)\qquad (f\,\hat{*}\,g)(t)=\int_0^t du\int_0^u f(u-\tau)g(\tau)d\tau+\Phi(f)\Phi(g)$$

$$+\Phi_x\left\{\int_t^x f(x+t-\tau)g(\tau)d\tau\right\}+\Phi(f)lg(t)+\Phi(g)lf(t),$$

where Φ is a linear functional in $\mathscr{C}(\Delta)$, and l is the Volterra integration operator, is a convolution of the integration operator $Lf=lf+\Phi(f)$ in $\mathscr{C}(\Delta)$, such that the representation

(38) $$L^2f=\{1\}\widehat{*}f$$

holds.

Proof. We shall show that if $f, g\in\mathscr{C}^1(\Delta)$, then $f\widehat{*}g=L(f*g)$. In fact denoting $h(t, x)=\int_t^x f(x+t-\tau)g(\tau)d\tau$, we have

$$\frac{\partial h}{\partial x}=\frac{\partial h}{\partial t}+f(x)g(t)+f(t)g(x)$$

and hence

$$f*g(t)=\int_0^t f(t-\tau)g(\tau)d\tau+\frac{d}{dt}\Phi_x\{h(t,x)\}+\Phi(f)g(t)+\Phi(g)f(t).$$

Therefore

$$L(f*g)=l\left\{\int_0^t f(t-\tau)g(\tau)d\tau\right\}+\Phi_x\{h(t,x)\}-\Phi_x\{h(0, x)\}$$

$$+\Phi(f)lg+\Phi(g)lf+\Phi_x\left\{\int_0^x f(x-\tau)g(\tau)d\tau\right\}+\Phi_x\Phi_y\left\{\frac{\partial h(y,x)}{\partial x}\right\}.$$

Let us simplify the last term. Since

$$\frac{\partial h(y, x)}{\partial x}=f(y)\,g(x)+\int_y^x f'(x+y-\tau)g(\tau)d\tau,$$

then

$$\Phi_y\Phi_x\left\{\frac{\partial h(y, x)}{\partial x}\right\}=\Phi_y\{f(y)\}\Phi_x\{g(x)\}=\Phi(f)\Phi(g),$$

according to Lemma 1.

Thus we have proved that $f\widehat{*}g=L(f*g)$ in $\mathscr{C}^1(\Delta)$. Hence the operation $\widehat{*}$ is a convolution of L in $\mathscr{C}^1(\Delta)$, since $Lf=\{1\}*f$. But since the operation (37) is continuous in $\mathscr{C}(\Delta)$, it is a bilinear, commutative and associative operation in $\mathscr{C}(\Delta)$, and hence it is a convolution of L not only in $\mathscr{C}^1(\Delta)$, but in $\mathscr{C}(\Delta)$ too. Representation (38) is fulfilled in the whole $\mathscr{C}(\Delta)$, as it can be seen either directly or by means of approximation. □

Now we shall consider some classes of integration operators, for which there are simpler convolutions in $\mathscr{C}(\Delta)$. An important particular case is obtained when the defining projector F of the right inverse operator L of the differentiation is defined not only in $\mathscr{C}^1(\Delta)$ but in $\mathscr{C}(\Delta)$ foo. Such is e. g. the Volterra integration operator $lf(t)=\int_0^t f(\tau)d\tau$ for which $Ff(t)=f(0)$.

Theorem 2.1.6. (Berg [2]). *If F: $\mathscr{C}(\Delta)\to\mathbf{C}$ is a linear functional with $F\{1\}=1$, then the operation*

(39) $$(f*g)(t)=F_x\left\{\int_x^t f(x+t-\tau)g(\tau)d\tau\right\}$$

is a convolution of the integration operator

(40) $$L\,f(t)=\int_0^t f(\tau)d\tau-F_x\left\{\int_0^x f(\tau)d\tau\right\}$$

in $\mathscr{C}(\Delta)$, *such that* $Lf=\{1\}*f$.

Proof. According to Theorem 2.1.1, operation (39) is bilinear, commutative and associative in $\mathscr{C}(\Delta)$. In order to assert that it is a convolution of L in $\mathscr{C}(\Delta)$, it is sufficient to prove the representation formula $Lf=\{1\}*f$. This is easily checked.

Let us note that in this case the integration operator Lf for $f\in\mathscr{C}(\Delta)$ is defined as the solution $y=Lf$ of the elementary boundary value problem

(41) $$y'=f, \quad Fy=0.$$

Example. Let $Ff(t)=\frac{f(0)+f(1)}{2}$ in $\mathscr{C}(\Delta)$ with $[0,\ 1]\subset\Delta$. Obviously $F\{1\}=1$. The operator

(42) $$L\,f(t)=\int_0^t f(\tau)d\tau-\frac{1}{2}\int_0^1 f(\tau)d\tau,$$

obtained from formula (40), is appropriately to be named the Euler integration operator, since $L^n\{1\}=\frac{1}{n!}E_n(t)$ for a positive integer n, where $E_n(t)$ is the n-th Euler polynomial. According to the theorem, the operation

(43) $$(f*g)(t)=\frac{1}{2}\left[\int_0^t f(t-\tau)g(\tau)d\tau-\int_t^1 f(1+t-\tau)g(\tau)d\tau\right]$$

is a convolution of the *Euler integration operator* (42) in $\mathscr{C}(\Delta)$. □

By means of this convolution, we can write down at once a corresponding Taylor formula with a remainder term in the form of convolution. According to Lemma 4 from 1.3.4, for an arbitrary right inverse operator of the differentiation, we can write

$$f(t)=\sum_{k=0}^{n-1}L^kFD^kf(t)+L^nf^{(n)}(t).$$

Since in our case $L^nf^{(n)}=(L^{n-1}\{1\})*f^{(n)}$ and $L^k\{1\}=\frac{1}{k!}E_k(t)$, we get

(44) $$f(t)=\sum_{k=0}^{n-1}\frac{f^{(k)}(0)+f^{(k)}(1)}{2.\ k!}E_k(t)+\frac{1}{(n-1)!}E_{n-1}(t)*f^{(n)}(t),$$

which is exactly *Boole's formula* (see Nörlund [82], p. 34). If we introduce the denotation $\bar{E}_{n-1}(t)$ for the anti-periodic continuation of $E_{n-1}(t)$ outside the segment [0, 1], we can easily give the remainder term the form

$$R_n(f;t)=\frac{1}{2(n-1)!}\int_0^1 \bar{E}_{n-1}(t-\tau)\,f^{(n)}(\tau)d\tau.$$

Let us now consider another general class of integration operators which have convolutions with a unit in $\mathscr{C}(\Delta)$.

Theorem 2.1.7. *If a defining projector F of an integration operator L of d/dt in $\mathscr{C}(\Delta)$ is defined in the whole space $\mathscr{C}(\Delta)$ and admits a representation of the form*

$$F(f)=\chi(lf) \tag{45}$$

with a continuous linear functional χ in $\mathscr{C}(\Delta)$, then the operation

$$(f\tilde{*}g)(t)=\chi_x\left\{\int_x^t f(t+x-\tau)\,g(\tau)d\tau\right\} \tag{46}$$

$$-\chi\{1\}\int_0^t f(t-\tau)g(\tau)d\tau+f(t)\chi(lg)+g(t)\chi(lf)$$

is a convolution of the operator $Lf=lf-\chi(l^2f)$ in $\mathscr{C}(\Delta)$, such that $\{1\}*f=f$.

Proof. The assertion follows immediately from Lemma 5 and from the easily verifiable identities

$$\{1\}*f=f \quad\text{and}\quad Lf=(t-\chi(\{x^2/2\}))*f.$$

Example. The right inverse operator of the differentiation

$$Lf(t)=\int_0^t f(\tau)d\tau-\int_0^1(1-\tau)\,f(\tau)d\tau$$

has a defining projector

$$Ff(t)=\int_0^1 f(\tau)d\tau$$

of the form considered. Indeed, we have $Ff(t)=\int_0^1 f(\tau)d\tau=\chi(lf)$, where $\chi(f)=f(1)$. Hence the operation

$$(f*g)(t)=f(t)\int_0^1 g(\tau)d\tau+g(t)\int_0^1 f(\tau)d\tau$$

$$-\int_0^t f(t-\tau)g(\tau)d\tau - \int_t^1 f(1+t-\tau)g(\tau)d\tau$$

is a convolution of L in $\mathscr{C}([0, 1])$, such that $\{1\}*f=f$. □

Later on we shall consider in greater detail this integration operator. It is closely connected with the Bernoulli polynomials. In fact, for each integer n, $n\geqq 0$, we have

$$L^n\{1\}=\frac{1}{n!}B_n(t),$$

where $B_n(t)$ is the n-th *Bernoulli polynomial*. In this case the generalized Taylor formula

$$f=\sum_{k=0}^{n-1} L^k FD^k f+L^nD^n f$$

takes the form

$$f(t)=\int_0^1 f(\tau)d\tau+\sum_{k=1}^{n-1}[f^{(k-1)}(1)-f^{(k-1)}(0)]\frac{B_k(t)}{k!}+\frac{1}{n!}B_n*f^{(n)}.$$

If we denote by $\bar{B}_n(t)$ the periodic continuation of $B_n(t)$ outside the interval $[0, 1]$, we get the well-known generalized Taylor formula of the form

$$f(t)=\int_0^1 f(\tau)d\tau+\sum_{k=1}^{n}[f^{(k-1)}(1)-f^{(k-1)}(0)]\frac{B_k(t)}{k!}-\frac{1}{n!}\int_0^1 \bar{B}_n(t-\tau)f^{(n)}(\tau)d\tau.$$

It is known as the *Euler-Maclaurin formula*. Having at our disposal the above convolution, we easily obtained the remainder term of the Euler-Maclaurin formula.

Finally, let us consider the right inverse operator of the differentiation operator, corresponding to a given generating function $E(\lambda)$ with $E(0)=1$ in the space $\mathscr{H}(\bar{D})$ of the locally holomorphic functions on a closed convex domain $\bar{D}$, containing all the singularities of the Borel transform $\gamma(t)$ of $E(\lambda)$. As we have seen, this integration operator has the form

$$Lf(z)=\int_0^z f(\zeta)d\zeta-\frac{1}{2\pi i}\int_{\Gamma_f}\gamma(\zeta)d\zeta\int_0^\zeta f(\sigma)d\sigma, \tag{47}$$

where it is assumed that $0\in D$.

Theorem 2.1.8. *The operation*

$$(f*g)(z)=\frac{1}{2\pi i}\int_\Gamma \gamma(\zeta)\left\{\int_\zeta^z f(z+\zeta-\sigma)g(\sigma)d\sigma\right\}d\zeta, \tag{48}$$

where Γ is a contour lying in the domains of the representatives $f(z)$ and $g(z)$ of the corresponding locally holomorphic functions, and con-

taining $\bar{D}$ in its inside, is a convolution of the integration operator (47) *in $\mathscr{H}(\bar{D})$, such that $Lf=\{1\}*f$.*

Proof. The bilinearity, commutativity and associativity of (48) can be proved in the same way as those of the operation in Theorem 2.1.1, but instead of the Weierstrass approximation theorem we should use the Runge approximation theorem. Then, in order to prove that (48) is a convolution of L in $\mathscr{H}(\bar{D})$, it is enough to verify the identity $Lf=\{1\}*f$. □

2.1.3. Convolutional representations of the commutants of linear integration operators. Here we shall use the convolutions found in the previous sections for obtaining explicit representations of the commutants of linear integration operators in $\mathscr{C}(\Delta)$, $\mathscr{L}(\Delta)$ and $\mathscr{H}(\bar{D})$.

In 1.1.3 we have found such representations for the Volterra integration operator by means of the Duhamel convolution. Now we shall use the results of 1.3.5. According to Theorem 1.3.11 for an annihilators-free and continuous convolution of a linear operator with a cyclic element in a Fréchet space, the commutant of the operator coincides with the multiplier ring of the corresponding convolution. Since all the convolutions of integration operators found in this chapter are annihilators-free, and since each integration operator of the form

$$Lf(t)=\int_0^t f(\tau)d\tau+\Phi_x\left\{\int_0^x f(\tau)\,d\tau\right\} \tag{49}$$

(49) has as a cyclic element the constant function $k=\{1\}$, then this theorem is applicable to integration operators.

As a first task we shall characterize a broad class of cyclic elements of (49) in $\mathscr{C}(\Delta)$.

Theorem 2.1.9 (Dimovski [43]). *A function $k\in\mathscr{C}^2(\Delta)$ is a cyclic element of operator* (47) *iff it is not a divisor of zero of its convolution* (33) *and if its projector $Ff(x)=f(0)-\Phi(f')$ does not vanish on it,* i. e. $Fk\neq 0$.

Proof. It is easy to prove the necessity of these two conditions. If $k\in\mathscr{C}^2(\Delta)$ is a divisor of zero of $*$, then $k*g=0$ for a $g\in\mathscr{C}^1(\Delta)$ which is not the zero function $\{0\}$. Then $(L^n k)*g=0$ for $n=0, 1, 2,\ldots$. If k were a cyclic element for L, then we would have $f*g=0$ for each $f\in\mathscr{C}(\Delta)$. But for $f=1$ we get a contradiction.

Also, if $F(k)=k(0)-\Phi(k')=0$, then $FL^n k=0$ for $n=0, 1, 2,\ldots$. Therefore, by means of linear combinations of $L^n k$ we could approximate only functions $f\in\mathscr{C}^1(\Delta)$ which have the property $Ff=0$. By such combinations we could not approximate e. g. the function $\{1\}$.

Let us prove the sufficiency of the hypothesis. If $f\in\mathscr{C}(\Delta)$ is arbitrarily chosen, then we search a function $g\in\mathscr{C}(\Delta)$, such that

$$F(k)g+k'*g=f, \tag{50}$$

where by $*$ the operation (33) is denoted. Equation (50) is a Fredholm integral equation of the second kind. As it is well known, according to the *Fredholm alternative,* for the existence of a solution g of (50) it is sufficient the homogeneous equation

(51) $$F(k)x+k'*x=0$$

to have the trivial solution $x=0$ only. Let x be an arbitrary solution of (51) from $\mathscr{C}(\Delta)$. Applying the operator L to both sides of (51), and taking into account the representation

(52) $$Lk'+F(k)=k,$$

we get $k*x=0$. But by the hypothesis k is a non-zero non-divisor of zero of convolution (33). Hence $x=0$. Thus the existence of such solution of (50) is ensured.

Denote $A_n(t)=L^n\{1\}$, $n=0, 1, 2, \ldots$. This is a set of Appell polynomials. We have

(53) $$L^nk=F(k)\{A_n(t)\}+k'*\{A_n(t)\}, \quad n=0, 1, 2, \ldots.$$

Then, choosing a polynomial sequence $\{g_n(t)\}_{n=0}^{\infty}$, which converges almost uniformly to g in $\mathscr{C}(\Delta)$, each $g_n(t)$ can be represented in the form

$$g_n(t)=\sum_{j=0}^{n} a_{nj}A_j(t), \quad n=0, 1, 2, \ldots .$$

Then, from (53) and (50) it follows that the sequence

$$f_n=\sum_{j=0}^{n} a_{nj}L^jk=F(k)g_n+k'*g_n$$

converges almost uniformly to f in $\mathscr{C}(\Delta)$. Hence k is a cyclic element of the operator L in $\mathscr{C}(\Delta)$. □

Theorem 2.1.10. *If M: $\mathscr{C}(\Delta)\to\mathscr{C}(\Delta)$ is a linear operator commuting with the integration operator $Lf=lf+\Phi(f)$, then M admits a representation of the form*

(54) $$Mf(t)=\frac{d^2}{dt^2}(m\mathbin{\widehat{*}}f),$$

where by $\widehat{*}$ *the convolution* (37) *is denoted, and* $m=M\{1\}$.

Proof. According to Theorem 1.3.11, each such operator M is a multiplier of the convolution (37). Then, applying the operator M to the identity $L^2f=\{1\}\mathbin{\widehat{*}}f$, we get $ML^2f=(M\{1\})\mathbin{\widehat{*}}f$. But $ML^2=L^2M$, and hence

$$L^2Mf=m\mathbin{\widehat{*}}f.$$

Applying d^2/dt^2 to the last identity, we get (54) with $m=M\{1\}$. □

Corollary 1. A linear operator M: $\mathscr{C}(\Delta)\to\mathscr{C}(\Delta)$ with $M\{1\}=m(t)\in\mathscr{C}^2(\Delta)$ commutes with the integration operator $Lf=lf+\Phi(f)$ in $\mathscr{C}(\Delta)$ iff it admits a representation of the form

(55) $$Mf=F(m)f+m'*f,$$

where by $*$ the operation (33) is denoted, and $F(m)=m(0)-\Phi(m')$.

Proof. If $m\in\mathscr{C}^2(\Delta)$, then the expression for $m*f$ is defined for each $f\in\mathscr{C}(\Delta)$, and

(56) $$m \widehat{*} f = L(m * f).$$

Then representation (54) gives

(57) $$Mf = \frac{d}{dt}(m * f) = F(m)f + F_x\left\{\int_x^t m'(t+x-\tau) f(\tau)d\tau\right\},$$

which is nothing but representation (55). In this we have used the notation $Ff = f(0) - \Phi(f')$ for the defining projector of L. □

Corollary 2. A linear operator $M: \mathscr{C}(\Delta) \to \mathscr{C}(\Delta)$ with $M\{1\} = m(t) \in C^1(\Delta)$ commutes with the integration operator $Lf = lf - F(lf)$ with a linear functional F in $\mathscr{C}(\Delta)$ with $F\{1\} = 1$, iff M has a representation of the form (55), where $*$ is operation (39).

This is shown in the same way as in the previous corollary.

Corollary 3. A linear operator M: $\mathscr{C}(\Delta) \to \mathscr{C}(\Delta)$ commutes with the integration operator $Lf = lf - \chi(l^2 f)$ in $\mathscr{C}(\Delta)$, where χ is a linear functional in $\mathscr{C}(\Delta)$ with $\chi(\{t\}) = 1$ iff it admits a representation of the form

(58) $$Mf = m \tilde{*} f$$

with $m = M\{1\}$, and with operation (46) as $\tilde{*}$.

Proof. Since $f = \{1\} \tilde{*} f$, then $Mf = M(\{1\} \tilde{*} f) = (M\{1\} \tilde{*} f$. This can be written due to the fact that M is a multiplier of the convolution $\tilde{*}$.

Finally, let us consider the integration operators in the space $\mathscr{H}(\bar{D})$ of locally holomorphic functions $f(z)$ on a finite closed convex domain $\bar{D}$, which contains all the singularities of the Borel transform $\gamma(z)$ of a given entire function $E(\lambda)$ of exponential type with $E(0) = 1$. In this case we can give a complete characterization of the commutant of the general integration operator

(59) $$Lf(z) = \int_0^z f(\zeta)d\zeta - \frac{1}{2\pi i}\int_\Gamma \gamma(\zeta)d\zeta \int_0^\zeta f(\sigma)d\sigma$$

in $\mathscr{H}(\bar{D})$. □

Theorem 2.1.11. *A linear operator M: $\mathscr{H}(\bar{D}) \to \mathscr{H}(\bar{D})$ commutes with the integration operator* (59) *iff it admits a representation of the form*

(60) $$Mf(z) = \mu f(z) + m * f(z)$$

with $\mu =$ const, and $m(z) = (M\{1\})'$, where by $*$ the operation (48) *is denoted. This representation is uniquely determined.*

Proof. By Theorem 2.1.8, the operation

$$(f * g)(z) = \frac{1}{2\pi i}\int_\Gamma \gamma(\zeta)\left(\int_\zeta^z f(z+\zeta-\sigma)g(\sigma)d\sigma\right)d\zeta$$

is a convolution of the operator L in $\mathscr{H}(\bar{D})$ with $Lf = \{1\} * f$. Since $\{1\}$ is again a cyclic element of L in $\mathscr{H}(\bar{D})$, then the set of the multipliers of this convolution coincides with the commutant of L. If $M: \mathcal{H}(\bar{D})\to\mathcal{H}(\bar{D})$ is a linear operator, commuting with L, let us apply it to the identity, $Lf=\{1\}*f$. Since M is a multiplier of the convolution $*$, then $MLf=(M\{1\})*f$ hence $L(Mf)=(M\{1\})*f$, since $ML=LM$. It remains to differentiate the last identity and to use the obvious identity

$$(f*g)'=F(f)g+f'*g$$

with $F(f)=(1/2\pi i)\int\limits_{\Gamma}\gamma(z)f(z)dz$ in order to obtain (60).

Conversely, it is clear that each operator of the form (60) commutes with the convolutional operator $Lf=\{1\}*f$. □

The results concerning the commutants of the integration operators in $\mathscr{C}(\Delta)$ can easily be transferred to the space $\mathscr{L}(\Delta)$ of locally summable functions on $\mathscr{C}(\Delta)$. This can be done by the following modifications of Theorems 2.1.4, 2.1.5 and 2.1.6.

Theorem 2.1.4′. *If Φ is an arbitrary linear functional in $\mathscr{L}(\Delta)$, then operation* (33) *is a convolution of the integration operator $Lf=lf+\Phi(f)$ in the space $\mathscr{AC}(\Delta)$ of the locally absolutely continuous functions on Δ, and $Lf=\{1\}*f$.*

For a proof one can use Lemmas 3 and 4 from sect. 2.1.

Theorem 2.1.5′. *If Φ is an arbitrary linear functional in $\mathscr{L}(\Delta)$, then operation* (37) *is a convolution of the integration operator $Lf=lf+\Phi(f)$ in $\mathscr{L}(\Delta)$, such that $L^2f=\{1\}*f$.*

Theorem 2.1.6′. *If F is a linear functional in $\mathscr{L}(\Delta)$ with $F\{1\}=1$, then operation* (40) *is a convolution of the integration operator $Lf=lf-F(lf)$ in $\mathscr{L}(\Delta)$, such that the representation $Lf=\{1\}*f$ holds.*

Then Theorem 2.1.10 and all its three corollaries remain valid with evident modifications, e. g. instead of the requirement $m\in\mathscr{C}^2(\Delta)$ in Corollary 1, we should assume that $m\in\mathscr{AC}^1(\Delta)$, i. e. that m is differentiable and that its derivative is locally absolutely continuous.

2.1.4. The commutant of the differentiation operator in an invariant hyperplane. We shall use the results obtained in the previous section to solve a problem in which there is no mention about convolutions and right inverses of the differentiation, but its solution requires use of these notions.

In Corollary 2 of Theorem 2.1.10 we have shown that a linear operator $M:\mathscr{C}(\Delta)\to\mathscr{C}(\Delta)$ with $M\{1\}=m\in\mathscr{C}^1(\Delta)$ commutes with the integration operator $Lf=lf-F(lf)$ with a linear functional F in $\mathscr{C}(\Delta)$ with $F\{1\}=1$, iff M admits a representation of the form

$$Mf(z)=F(m)f(z)+F_x\left\{\int\limits_x^t m'(x+t-\tau)f(\tau)d\tau\right\}. \tag{61}$$

In this case it is easy to verify that $F(Mf)=0$ for $f\in\mathscr{C}(\Delta)$, provided $F(f)=0$. In ther words, the hyperplane $F(f)=0$ is an invariant subspace of the operator M. Let $f\in\mathscr{C}(\Delta)$ and let f lie in this hyperplane, i. e. $F(f)=0$. Then $f=F(f)+Lf'=Lf'$. Therefore $Mf=MLf'=LMf'$ and hence

$$\frac{d}{dt}Mf=M\frac{d}{dt}f. \tag{62}$$

We have seen that the operator M commutes with the differentiation operator in its invariant hyperplane.

This suggests an inverse problem: If $\Phi: \mathscr{C}(\Delta)\to\mathbf{C}$ is an arbitrary non-zero linear functional in $\mathscr{C}(\Delta)$, find all linear operators $M: \mathscr{C}(\Delta)\to\mathscr{C}(\Delta)$, which map smooth functions in smooth ones, i. e. $M: \mathscr{C}^1(\Delta)\to\mathscr{C}^1(\Delta)$ such that the hyperplane $\mathscr{C}_\Phi=\{f\in\mathscr{C}(\Delta): \Phi(f)=0\}$ is an invariant subspace for M, and which commute with the differentiation operator d/dt in this hyperplane. In other words, we ask for fulfilling of (62) for each $f\in\mathscr{C}^1_\Phi$.

Theorem 2.1.12 (Dimovski [38]). *A linear operator $M: \mathscr{C}(\Delta)\to\mathscr{C}(\Delta)$, such that $M: \mathscr{C}^1(\Delta)\to\mathscr{C}^1(\Delta)$ and having the hyperplane $\mathscr{C}_\Phi=\{f\in\mathscr{C}(\Delta): \Phi(f)=0\}$ as an invariant subspace, where Φ is a non-zero linear functional in $\mathscr{C}(\Delta)$, commutes with the differentiation operator d/dt in $\mathscr{C}_\Phi$ iff it admits a representation of the form*

$$Mf(t)=\mu f(t)+\Phi_x\left\{\int_x^t m(t+x-\tau)f(\tau)d\tau\right\} \tag{63}$$

with $\mu=$const *and with* $m(t)\in\mathscr{C}(\Delta)$.

Proof. By Theorem 2.1.1, the operation

$$(f*g)(t)=\Phi_x\left\{\int_x^t f(t+x-\tau)g(\tau)d\tau\right\} \tag{64}$$

is a bilinear, commutative and associative operation in $\mathscr{C}(\Delta)$. Moreover, (64) is annihilators-free in $\mathscr{C}(\Delta)$. By Lemma 1, for arbitrary $f, g\in\mathscr{C}(\Delta)$ we have $\Phi(f*g)=0$. Hence $*: \mathscr{C}(\Delta)\times\mathscr{C}(\Delta)\to\mathscr{C}_\Phi$. Therefore, every operator of the form (63) has Φ as an invariant subspace. If $f\in\mathscr{C}^1_\Phi$, then from (63) we get

$$(Mf)'=\mu f'+\Phi_x\left\{\int_x^t m(t+x-\tau)f'(\tau)d\tau+m(t)f(x)\right\},$$

i. e. $(Mf)'=Mf'$. Hence, each operator of the form (63) commutes with d/dt in $\mathscr{C}_\Phi$.

Let now $M: \mathscr{C}(\Delta)\to\mathscr{C}(\Delta)$ be an arbitrary linear operator, having the hyperplane $\mathscr{C}_\Phi$ as an invariant subspace and such that $M: \mathscr{C}^1(\Delta)\to\mathscr{C}^1(\Delta)$. We shall show that M admits a representation of the form (63). In the proof of Theorem 2.1.1 we have seen that if $\lambda\in\mathbf{C}$ is such that $E(\lambda)=\Phi_x\{e^{\lambda x}\}\neq0$, then operation (64) is a convolution of the right inverse operator L_λ of the differential operator $D_\lambda=(d/dt)-\lambda$, given by

$$L_\lambda f(t)=e^{\lambda t}\int_0^t e^{-\lambda\tau}f(\tau)d\tau-\frac{e^{\lambda t}}{E(\lambda)}\Phi_x\left\{e^{\lambda x}\int_0^x e^{-\lambda\tau}f(\tau)d\tau\right\}. \tag{65}$$

The function $e_\lambda=\{e^{\lambda t}\}$ is a cyclic element of L_λ in $\mathscr{C}(\Delta)$. Indeed, we have $L^n e_\lambda=\{e^{\lambda t}P_n(t)\}$, where $P_n(t)$ is a polynomial of exactly n-th degree. Then, by Theorem 1.3.11, the operator L_λ is a multiplier of the convolution (64) in $\mathscr{C}(\Delta)$. From Corollary 2 of the same theorem, or from (66), we get

(66)
$$Mf(t)=\left(\frac{d}{dt}-\lambda\right)(n*f) \quad \text{with} \quad n=M\left\{\frac{e^{\lambda t}}{E(\lambda)}\right\}.$$

But this is a representation of the form (63) with $\mu=\Phi\{n\}$ and $m=D_\lambda n$. The theorem is proved. □

Open problem. Find a representation of the linear operators $M:\mathscr{C}(\Delta)\to\mathscr{C}(\Delta)$ with an invariant hyperplane $\mathscr{C}_\Phi$ and commuting with the differentiation operator in $\mathscr{C}^1_\Phi$, without the assumption that M maps necessarily smooth functions into smooth ones.

2.2. AN APPLICATION OF THE CONVOLUTIONAL APPROACH TO DIRICHLET EXPANSIONS OF LOCALLY HOLOMORPHIC FUNCTIONS

In the previous section we have found convolutions of the linear right inverse operators of the differentiation operator in the space $\mathscr{H}(\bar{D})$ of the *locally holomorphic functions* in a finite closed convex domain $\bar{D}$, and we characterized completely their commutants (Theorems 2.1.3, 2.1.8 and 2.1.11). Let us now summarize the corresponding results. If $E(\lambda)$ is a given entire function of exponential type with $E(0)=1$, and if $\gamma(z)$ is its Borel transform, then let $\bar{D}$ be the smallest closed convex domain in the complex plane, which contains all the singularities of $\gamma(z)$. We have shown that the right inverse operator

(1)
$$Lf(z)=\int_0^z f(\zeta)d\zeta-\frac{1}{2\pi i}\int_\Gamma \gamma(\zeta)d\zeta\int_0^\zeta f(\sigma)d\sigma$$

of the differentiation operator d/dz in $\mathscr{H}(\bar{D})$ has as a defining projector the linear functional

(2)
$$Ff(z)=f(z)-Lf'(z)=\frac{1}{2\pi i}\int_\Gamma \gamma(z)f(z)dz.$$

Then the indicatrix $F_z\{e^{\lambda z}\}$ of the functional F coincides with the given entire function $E(\lambda)$, i. e. $E(\lambda)=F_z\{e^{\lambda z}\}$.

The operation

(3)
$$(f*g)(z)=\frac{1}{2\pi i}\int_\Gamma \gamma(\zeta)d\zeta\int_\zeta^z f(z+\zeta-\sigma)g(\sigma)d\sigma$$

with a contour Γ, contained in the domains of f and g and containing $\bar{D}$, is a convolution of L in $\mathscr{H}(\bar{D})$, such that L has a representation as the convolutional operator $Lf=\{1\}*f$.

The commutant of L in $\mathscr{H}(\bar{D})$ coincides with the multiplier ring of convolution (3), and each linear operator $M:\mathscr{H}(\bar{D})\to\mathscr{H}(\bar{D})$ which commutes with L in $\mathscr{H}(\bar{D})$, has the representation

(4)
$$Mf=\mu f+m*f$$

with $\mu=\text{const}$, and with $m\in\mathscr{H}(\bar{D})$. This representation is uniquely determined.

All these results can immediately be applied to the problem of expanding of locally holomorphic functions $f(z)$ on a given system of exponents $e^{\lambda_k z}$, where $\lambda_1, \lambda_2, \ldots$ are the zeros of an entire function of exponential type. The investigations in this direction during the recent time are accounted in the book of Leontiev [73]. Here we propose an alternative algebraic approach, related to Hromov's approach [64].

2.2.1. Delsarte-Leontiev formulas for the coefficients of Dirichlet expansions. Let $E(\lambda)$ be an entire function of exponential type with $E(0)=1$. First, we assume that $E(\lambda)$ has infinitely many simple zeros λ_k, $k=1, 2, \ldots$. By $\gamma(z)$ we denote the Borel transform of $E(\lambda)$, and by $\bar{D}$ the smallest closed convex domain, containing all the singularities of $\gamma(z)$. As usual, we identify the elements of $\mathscr{H}(\bar{D})$ with their representatives. This shall not cause any confusion. Assume that a function $f(z) \in \mathscr{H}(\bar{D})$ is developable in a uniformly convergent Dirichlet series of the form

(5) $$f(z)=\sum_{k=1}^{\infty} a_k e^{\lambda_k z}$$

in a domain $G \supset \bar{D}$. We shall show that in this case the coefficients a_k are uniquely determined.

Lemma 1. *The exponents $e^{\lambda_k z}$, $k=1, 2, \ldots$, are the eigenfunctions of the right inverse operator* (1) *of the differentiation operator d/dz in $\mathscr{H}(\bar{D})$.*

Proof. We have

$$Le^{\lambda_k z}=\frac{1}{\lambda_k} e^{\lambda_k z}-\frac{1}{2\pi i}\int_\Gamma \gamma(\zeta)\frac{1}{\lambda_k} e^{\lambda_k \zeta} d\zeta$$

since $\frac{1}{2\pi i}\int_\Gamma \gamma(\zeta)d\zeta=1$, due to the assumption $E(0)=1$. But

(6) $$E(\lambda)=\frac{1}{2\pi i}\int_\Gamma \gamma(\zeta)e^{\lambda\zeta}d\zeta$$

and hence $Le^{\lambda_k z}=\frac{1}{\lambda_k}e^{\lambda_k z}$. Therefore $e^{\lambda_k z}$ is an eigenfunction of L with the eigenvalue $1/\lambda_k$. □

Lemma 2. *If $f * g$ is the convolution* (3), *then*

(7) $$\{e^{\lambda_j z}\} * \{e^{\lambda_k z}\}=\begin{cases} 0 & \text{if } j \neq k \\ -E'(\lambda_k)e^{\lambda_k z} & \text{if } j=k. \end{cases}$$

Proof. Using representation (6), for $E(\lambda)$ we get

$$\{e^{\lambda_j z}\} * \{e^{\lambda_k z}\}=\frac{e^{\lambda_j z}}{2\pi i}\int_\Gamma \gamma(\zeta)e^{\lambda_j \zeta}d\zeta \int_\zeta^z e^{(\lambda_k-\lambda_j)\sigma}d\sigma$$

$$=\frac{e^{\lambda_j z}}{2\pi i(\lambda_k-\lambda_j)}\int_{\Gamma}\gamma(\zeta)e^{\lambda_j\zeta}\,[e^{(\lambda_k-\lambda_j)z}-e^{(\lambda_k-\lambda_j)\zeta}]d\zeta$$

$$=\frac{1}{\lambda_k-\lambda_j}\,[E(\lambda_j)e^{\lambda_k z}-E(\lambda_k)e^{\lambda_j z}]=0,$$

since λ_j and λ_k are zeros of $E(\lambda)$. For $j=k$ we have

$$\{e^{\lambda_k z}\}^{*2}=\frac{e^{\lambda_k z}}{2\pi i}\int_{\Gamma}\gamma(\zeta)\,(z-\zeta)\,e^{\lambda_k\zeta}d\zeta$$

$$=e^{\lambda_k z}[zE(\lambda_k)-E'(\lambda_k)]=-E'(\lambda_k)e^{\lambda_k z}.$$

This proves the lemma. □

Lemma 3. *If* $f(z)\in\mathscr{H}(\bar{D})$, *then*

(8) $$f*\{e^{\lambda_k z}\}=\chi_k(f)\,e^{\lambda_k z},$$

where $\chi_k(f)$ *is the linear functional*

(9) $$\chi_k(f)=-\frac{1}{2\pi i}\int_{\Gamma}\gamma(\zeta)d\zeta\int_0^{\zeta}e^{\lambda_k\tau}\,f(\zeta-\tau)d\tau.$$

Proof. By (3) we get

$$f*\{e^{\lambda_k z}\}=\frac{1}{2\pi i}\int_{\Gamma}\gamma(\zeta)d\zeta\int_{\zeta}^{z}e^{\lambda_k(z+\zeta-\sigma)}\,f(\sigma)d\sigma$$

$$=\frac{e^{\lambda_k z}}{2\pi i}\int_{\Gamma}e^{\lambda_k\zeta}\,\gamma(\zeta)d\zeta\int_{\zeta}^{z}e^{-\lambda_k\sigma}\,f(\sigma)d\sigma.$$

Using the obvious identity

$$\int_{\zeta}^{z}e^{-\lambda_k\sigma}f(\sigma)d\sigma=\int_0^{z}e^{-\lambda_k\sigma}f(\sigma)\,d\sigma-\int_0^{\zeta}e^{-\lambda_k\sigma}f(\sigma)d\sigma,$$

we get

$$f*\{e^{\lambda_k z}\}=e^{\lambda_k z}\,E(\lambda_k)\int_0^{z}e^{-\lambda_k\sigma}f(\sigma)d\sigma$$

$$-\frac{e^{\lambda_k z}}{2\pi i}\int_{\Gamma}\gamma(\zeta)d\zeta\int_0^{\zeta}e^{\lambda_k(\zeta-\sigma)}\,f(\sigma)d\sigma=-\frac{e^{\lambda_k z}}{2\pi i}\int_{\Gamma}\gamma(\zeta)d\zeta\int_0^{\zeta}e^{\lambda_k\tau}f(\zeta-\tau)d\tau.$$

This is just representation (8). □

Lemma 4. *If a function $f(z)\in\mathscr{H}(\bar{D})$ is developable in a uniformly convergent Dirichlet series* (5) *in a domain $G\supset\bar{D}$, then the coefficients a_k are uniquely determined and*

$$a_k=\frac{1}{2\pi iE'(\lambda_k)}\int_{\Gamma}\gamma(\zeta)d\zeta\int_0^{\zeta}e^{\lambda_k\tau}f(\zeta-\tau)d\tau. \tag{10}$$

Proof. Let us multiply convolutionally (5) by $e^{\lambda_k z}$. Using Lemmas 2 and 3, we get

$$-a_kE'(\lambda_k)e^{\lambda_k z}=-\frac{e^{\lambda_k z}}{2\pi i}\int_{\Gamma}\gamma(\zeta)d\zeta\int_0^{\zeta}e^{\lambda_k\tau}f(\zeta-\tau)d\tau,$$

which gives (10). □

Definition 1. The *Delsarte-Leontiev coefficients* of a locally holomorphic function $f(z)\in\mathscr{H}(\bar{D})$ with respect to the exponent system $\{e^{\lambda_k z}\}_{k=1}^{\infty}$ where λ_k are simple zeros of an entire function $E(\lambda)$ of exponential type with $E(0)=1$, are said to be the numbers

$$a_k(f)=\frac{1}{2\pi iE'(\lambda_k)}\int_{\Gamma}\gamma(\zeta)d\zeta\int_0^{\zeta}e^{\lambda_k\sigma}f(\zeta-\sigma)d\sigma. \tag{11}$$

Delsarte [27] was the first to introduce such functionals, but for real functions only. For the case of analytic functions these coefficients are introduced by A. F. Leontiev. To Leontiev [73], p. 255 is due the following uniqueness theorem:

If the entire function of exponential type $E(\lambda)$ has infinitely many and only simple zeros $\lambda_1, \lambda_2, \ldots$, then $a_k(f)=0$ for $k=1, 2, \ldots$ implies $f(z)\equiv 0$.

In fact this is only a particular case of a more general theorem due to Leontiev too. We shall state it later.

Following the analogy with the Fourier series, we give

Definition 2. The *formal Leontiev expansion* of a function $f(z)\in\mathscr{H}(\bar{D})$ on the exponents $\{e^{\lambda_k z}\}_{k=1}^{\infty}$, where λ_k are simple zeros of an entire function of exponential type $E(\lambda)$ with $E(0)=1$, is said to be the correspondence

$$f(z)\sim\sum_{k=1}^{\infty}a_k(f)e^{\lambda_k z} \tag{12}$$

with $a_k(f)$ given by (11).

The Leontiev expansion has the uniqueness property when $\lambda_1, \lambda_2, \ldots$, are all the zeros of $E(\lambda)$, and they are simple ones. Then one can consider the correspondence (11) as an integral transform in $\mathscr{H}(\bar{D})$ (see Dimovski [42]). Here we prefer to proceed in terms of the formal Leontiev expansions. Following the analogy with the Fourier series, here we can also consider the problem of developing the functions $f(z)\in\mathscr{H}(\bar{D})$ in a Dirichlet series on a given exponent system $\{e^{\lambda_k z}\}_{k=1}^{\infty}$. It is clear that without further restrictions on the functions $f(z)$ we could hardly expect a representation of the form

$$f(z)=\sum_{k=1}^{\infty} a_k(f)e^{\lambda_k z}$$

to be always valid in a domain $G\supset\overline{D}$, such that the series would be uniformly convergent in the closed subdomains of G. Indeed, then for a sufficiently small $t\in\mathbf{C}$ the function $f(z+t)$ is a function of $\mathscr{H}(\overline{D})$ too and it should be developable in a uniformly convergent series

(13)
$$f(z+t)=\sum_{k=1}^{\infty} a_k(f)e^{\lambda_k t}e^{\lambda_k z}$$

on the compact subdomains $G'\supset\overline{D}$. If we take a contour Γ, lying in the domain of $\varphi_t(z)=f(z+t)$ and containing $\overline{D}$ inside, then applying the functional

$$F(f)=\frac{1}{2\pi i}\int_{\gamma}\gamma(\zeta)f(\zeta)d\zeta$$

to both sides of (13), we get

(14)
$$M(t)=F_z\{f(z+t)\}=\frac{1}{2\pi i}\int_{\Gamma}\gamma(\zeta)f(\zeta+t)d\zeta\equiv 0$$

for every sufficiently small t. It is easy to figure out that the necessary condition (14) for developability of $f(z)$ is equivalent to

(15)
$$F(f^{(k)})=0,\quad k=0,\,1,\,2,\,\ldots\,.$$

That is why it is desirable to extend the above considerations to functions $f(z)$, which are holomorphic in the open domain D and are continuous on the closed domain $\overline{D}$. The space of all these functions will be denoted by $\dot{\mathscr{H}}(D)$. But in order to ensure the continuity of the Borel transform $\gamma(z)$ of $E(\lambda)$ on the boundary ∂D of D, it is necessary to impose further restrictions on the entire function $E(\lambda)$. Following Leontiev [73], p. 230, we subject $E(\lambda)$ to the additional restriction

(16)
$$|E(re^{i\varphi})|<A\frac{e^{h(\varphi)r}}{r^{\mu}}$$

for each $r>0$ with some $\mu>1$, where $h(\varphi)$ is the indicator function of $E(\lambda)$.

If (16) is fulfilled, and if D is the indicator diagram of $E(\lambda)$, then the Borel transform $\gamma(z)=\int_0^{\infty}e^{-\lambda z}E(\lambda)d\lambda$ of $E(\lambda)$ is regular outside $\overline{D}$ and continuous on ∂D.

Then the operation

(17)
$$(f*g)(z)=\frac{1}{2\pi i}\int_{\partial\overline{D}}\gamma(\zeta)\int_{\zeta}^{z}f(z+\zeta-\tau)g(\tau)d\tau$$

is a convolution of the integration operator

(18) $$Lf(z)=\int_0^z f(\zeta)d\zeta-\frac{1}{2\pi i}\int_{\partial D}\gamma(\zeta)d\zeta\int_0^\zeta f(\sigma)d\sigma$$

in $\overline{\mathscr{H}}(D)$, such that the representation $Lf=\{1\}*f$ holds.

Then the Delsarte-Leontiev coefficients of a function $f\in\overline{\mathscr{H}}(D)$ on the system $\{e^{\lambda_k z}\}_{k=1}^\infty$ of exponents, are named the linear functionals

(19) $$a_k(f)=\frac{1}{2\pi iE'(\lambda_k)}\int_{\partial D}\gamma(\zeta)d\zeta\int_0^\zeta e^{\lambda_k\sigma}f(\zeta-\sigma)d\sigma$$

in $\overline{\mathscr{H}}(D)$.

By the *second Leontiev uniqueness theorem* [73], p. 260, the coefficients $a_k(f)$ determine uniquely the function $f\in\overline{\mathscr{H}}(D)$, i. e. $a_k(f)=0$ for $k=1, 2,\ldots$, imply $f\equiv 0$.

Lemma 4. *If f, $g\in\overline{\mathscr{H}}(D)$, then*

(20) $$a_k(f*g)=-E'(\lambda_k)a_k(f)a_k(g),\quad k=1, 2,\ldots .$$

Proof. Since by definition

$$f*\{e^{\lambda_k z}\}=-E'(\lambda_k)a_k(f)e^{\lambda_k z},$$

then

$$(f*g)*\{e^{\lambda_k z}\}=-E'(\lambda_k)a_k(f)g*\{e^{\lambda_k z}\}=E'^2(\lambda_k)a_k(f)a_k(g)e^{\lambda_k z}$$

and hence

$$-E'(\lambda_k)a_k(f*g)=E'^2(\lambda_k)a_k(f)a_k(g). \ \square$$

Let us now consider the following Appell polynomials set

(21) $$A_n(z)=L^n\{1\}=\{1\}^{*(n+1)},\quad n=0, 1, 2,\ldots,$$

generated by operator (18). □

Lemma 5. *The formal Leontiev expansion of the polynomial $A_n(z)$, defined by* (21), *has the form*

(22) $$A_n(z)\sim-\sum_{k=0}^\infty\frac{e^{\lambda_k z}}{\lambda_k^{n+1}E'(\lambda_k)}\,.$$

Proof. Taking into account (20), it is enough to establish that

$$1\sim-\sum_{k=1}^\infty\frac{e^{\lambda_k z}}{\lambda_k E'(\lambda_k)}\,.$$

Indeed,

$$E'(\lambda_k)a_k(\{1\})=\frac{1}{2\pi i}\int_{\partial D}\gamma(\zeta)d\zeta\int_0^{\zeta}e^{\lambda_k\sigma}\,d\sigma$$

$$=\frac{1}{2\pi i\lambda_k}\int_{\partial \overline{D}}\gamma(\zeta)\,(e^{\lambda_k\zeta}-1)d\zeta=\frac{1}{\lambda_k}\,[E(\lambda_k)-1]=-\frac{1}{\lambda_k}\cdot\ \square$$

Now we can prove a theorem for developing an arbitrary function of $\overline{\mathscr{H}}(D)$ into a finite sum of polynomials and a uniformly convergent Dirichlet series on the given exponents system.

Theorem 2.2.1. *If for some $n>0$ the Leontiev expansion* (22) *of the Appell polynomial $A_n(z)$, given by* (21), *is uniformly convergent to $A_n(z)$ in $\overline{D}$, then each function $f(z)$ which is holomorphic on D and continuous with its first $n+1$ derivatives on $\overline{D}$, can be represented in the form*

$$f(z)=\sum_{j=0}^{n}F(f^{(j)})\,A_j(z)+\sum_{k=1}^{\infty}\frac{a_k(f^{(n+1)})e^{\lambda_k z}}{\lambda_k^{n+1}}, \tag{23}$$

where $a_k(f^{(n+1)})$ are the Delsarte-Leontiev coefficients of $f^{(n+1)}(z)$ and $F(\varphi)=\frac{1}{2\pi i}\int_{\partial D}\gamma(\zeta)\,\varphi(\zeta)\,d\zeta$. *The series is uniformly convergent on $\overline{D}$.*

Proof. The general Taylor formula from 1.3.4, applied to the operator L, gives

$$f(z)=\sum_{j=0}^{n}F(f^{(j)})A_j(z)+R_n(f;\ z)$$

with

$$R_n\{f,\ z\}=L^{n+1}f^{(n+1)}(z)=A_n(z)*f^{(n+1)}(z).$$

Under the hypothesis that

$$A_n(z)=-\sum_{k=1}^{\infty}\frac{e^{\lambda_k z}}{\lambda_k^{n+1}E'(\lambda_k)},$$

where the series is uniformly convergent on $\overline{D}$, we have

$$R_n(f,\ z)=-\sum_{k=1}^{\infty}\frac{f^{(n+1)}*e^{\lambda_k z}}{\lambda_k^{n+1}E'(\lambda_k)}=\sum_{k=1}^{\infty}\frac{a_k(f^{(n+1)})}{\lambda_k^{n+1}}\,e^{\lambda_k z}.$$

This series shall be uniformly convergent, since it is obtained by integration of an uniformly convergent series. $\square$

In a more general setting, the accelerating of the convergence of eigenexpansions of right inverse operators is considered in Tasche [100].

Corollary. Under the hypothesis of the theorem, each function $f(z)$ which is holomorphic in D and continuous along with its first $n+1$ derivatives on $\bar{D}$, can be developed in a uniformly convergent Leontiev expansion on $\bar{D}$, provided it satisfies the conditions

$$\frac{1}{2\pi i}\int_{\partial D}\gamma(\zeta)f^{(j)}(\zeta)d\zeta=0, \quad j=0, 1, 2, \ldots, n. \tag{24}$$

2.2.2. Convolutional representation of the multipliers of the formal Leontiev expansion. As in the previous section, let $E(\lambda)$ be an entire function of exponential type with only simple zeros $\lambda_1, \lambda_2, \ldots, \lambda_n, \ldots$ and let $\gamma(z)$ be its Borel transform and $\bar{D}$ be its indicator diagram. If $f\in\mathscr{H}(\bar{D})$, then

$$f(z)\sim\sum_{k=1}^{\infty} a_k(f)e^{\lambda_k z} \tag{25}$$

with

$$a_k=\frac{1}{2\pi iE'(\lambda_k)}\int_{\Gamma}\gamma(\zeta)d\zeta\int_0^{\zeta}e^{\lambda_k\sigma}f(\zeta-\sigma)d\sigma \tag{26}$$

is the formal Leontiev expansion on $f(z)$ on the exponents $\{e^{\lambda_k z}\}_{k=1}^{\infty}$.

Definition 3. A numerical sequence $\{\mu_k\}_{k=1}^{\infty}$ is said to be a *multiplier sequence* of the formal Leontiev expansion (25), iff for every $f\in\mathscr{H}(\bar{D})$ the series

$$\sum_{k=1}^{\infty}\mu_k a_k(f)e^{\lambda_k z} \tag{27}$$

is a formal Leontiev expansion on the same exponent system too.

In fact, each such multiplier sequence $\{\mu_k\}_{k=1}^{\infty}$ defines an operator M: $\mathscr{H}(\bar{D})\rightarrow\mathscr{H}(\bar{D})$, such that

$$Mf\sim\sum_{k=1}^{\infty}\mu_k a_k(f)e^{\lambda_k z}. \tag{28}$$

Such an operator M will be named a multiplier of the Leontiev expansion (25).

We are ready to describe all the multipliers of the Leontiev expansion (25) in $\mathscr{H}(\bar{D})$.

Theorem 2.2.2. *Each multiplier* M: $\mathscr{H}(\bar{D})\rightarrow\mathscr{H}(\bar{D})$ *of the Leontiev expansion* (25) *is a multiplier of the corresponding convolution* (3) *and it admits a representation of the form*

$$Mf(z)=\mu f(z)+(m*f)(z) \tag{29}$$

with $\mu=\text{const}$ *and with* $m\in\mathscr{H}(\bar{D})$. *Conversely, each operator of the form* (29) *is a multiplier of the Leontiev expansion* (25).

Proof. Let M: $\mathscr{H}(\bar{D})\rightarrow\mathscr{H}(\bar{D})$ be a multiplier of the formal Leontiev expansion (25). We shall show that $M(f*g)=(Mf)*g$ for arbitrary $f, g\in\mathscr{H}(\bar{D})$.

Let us determine the k-th Delsarte-Leontiev coefficient of the difference $M(f*g)-(Mf)*g$. We get

$$a_k[M(f*g)-(Mf)*g]=\mu_k a_k(f*g)+E'(\lambda_k)a_k(Mf)a_k(g)$$
$$=-\mu_k E'(\lambda_k)a_k(f)a_k(g)+E'(\lambda_k)\mu_k a_k(f)a_k(g)=0$$

for $k=0, 1, 2, \ldots$. From the Leontiev uniqueness theorem it follows that $M(f*g)-(Mf)*g=0$. By Theorem 2.1.11, the operator M has the representation (29) with $\mu=\text{const}$, and with $m\in\mathscr{H}(\overline{D})$. It remains to show that each operator of the form (29) is a multiplier of the formal Leontiev expansion (25). Indeed, using Lemma 4, we get

$$a_k(Mf)=\mu a_k(f)+a_k(m*f)=[\mu-E'(\lambda_k)a_k(m)]a_k(f),$$

i. e. M is a multiplier of the Leontiev expansion (25) with the multiplier sequence $\mu_k=\mu-E'(\lambda_k)a_k(m)$. □

Corollary. A sequence $\{\mu_k\}_{k=1}^{\infty}$ is a multiplier sequence of the Leontiev expansion (25) iff there exists a function $m\in\mathscr{H}(\overline{D})$ and a constant $\mu\in\mathbb{C}$, such that

$$\mu_k=\mu-\frac{1}{2\pi i}\int_{\Gamma}\gamma(\zeta)d\zeta\int_0^{\zeta}e^{\lambda_k\sigma}m(\zeta-\sigma)d\sigma,\quad k=1, 2, 3, \ldots. \tag{30}$$

2.2.3. Leontiev's expansions in the case of multiple zeros of the indicatrix. Let $E(\lambda)$ again be an entire function of exponential type, but its different zeros λ_k, which are supposed to be infinitely many, have corresponding multiplicities $m_k\geq 1$, $k=1, 2, \ldots$. In this case we use the same convolution

$$(f*g)(z)=\frac{1}{2\pi i}\int_{\Gamma}\gamma(\zeta)d\zeta\int_{\zeta}^{z}f(z+\zeta-\sigma)g(\sigma)d\sigma \tag{31}$$

of the integration operator

$$Lf(z)=\int_0^z f(\zeta)\,d\zeta-\frac{1}{2\pi i}\int_{\Gamma}\gamma(\zeta)d\zeta\int_0^{\zeta}f(\sigma)d\sigma \tag{32}$$

in $\mathscr{H}(\overline{D})$. In contrast to the case of a simple zero λ_k, for $m_k>1$, we have

$$\{e^{\lambda_k z}\}^{*2}=-E'(\lambda_k)e^{\lambda_k z}=0. \tag{33}$$

In this case we can use the results of 1.3.2. Then the corresponding eigenspace is not one-dimensional. We shall denote it by $\mathscr{E}_{\lambda_k}^{(m_k)}$.

The subspace

$$\mathscr{E}_{\lambda_k}^{(m_k)}=\ker(\lambda_k L-I)^{m_k} \tag{34}$$

of $\mathscr{H}(\overline{D})$ is m_k-dimensional and it consists of all quasi-polynomials of the form $Q_{m_k}(z)e^{\lambda_k z}$, where $Q_{m_k}(z)$ is a polynomial of at most (m_k-1)-th degree. In 1.3.2 we have shown that only from the assumption that the operator L has a convolution $*$ which has no annihilators in the space (34) the existence of a multiplier projector $P_\lambda f = u_\lambda * f$ on this subspace can be asserted. In our case an explicit formula for the corresponding projecting elements can be given.

Lemma 6. *Let C_k be a simple contour, containing in its inside the zero λ_k of $E(\lambda)$, and none of the other zeros of $E(\lambda)$. Then for*

$$\varphi_k(z) = -\frac{1}{2\pi i}\int_{C_k}\frac{e^{\lambda z}}{E(\lambda)}\,d\lambda \tag{35}$$

the identity

$$[\varphi_k(z)]^{*2} = \varphi_k(z) \tag{36}$$

holds.

Proof. Let C_k' and C_k'' be two simple contours, both of them containing λ_k, and no other zero of $E(\lambda)$, such that C_k'' is contained in the inside of C_k'. Expressing φ_k^{*2} as the convolutional product of

$$\varphi_k(z) = -\frac{1}{2\pi i}\int_{C_k'}\frac{e^{\lambda z}}{E(\lambda)}\,d\lambda,$$

and

$$\varphi_k(z) = -\frac{1}{2\pi i}\int_{C_k''}\frac{e^{\mu z}}{E(\mu)}\,d\mu,$$

we get

$$[\varphi_k(z)]^{*2} = \frac{1}{(2\pi i)^2}\int\int_{C_k'\times C_k''}\frac{e^{\lambda z} * e^{\mu z}}{E(\lambda)E(\mu)}\,d\lambda d\mu.$$

But

$$e^{\lambda z} * e^{\mu z} = \frac{E(\lambda)e^{\mu z} - E(\mu)e^{\lambda z}}{\mu-\lambda}$$

and therefore

$$[\varphi_k(z)]^{*2} = \frac{1}{(2\pi i)^2}\int_{C_k''}\frac{e^{\mu z}}{E(\mu)}\,d\mu\int_{C_k'}\frac{d\lambda}{\mu-\lambda}$$

$$-\frac{1}{(2\pi i)^2}\int_{C_k'}\frac{e^{\lambda z}}{E(\lambda)}\,d\lambda\int_{C_k''}\frac{d\mu}{\mu-\lambda} = -\frac{1}{2\pi i}\int_{C_k''}\frac{e^{\mu z}}{E(\mu)}\,d\mu = \varphi_k(z),$$

since

$$\int_{C_k'}\frac{d\lambda}{\mu-\lambda} = -2\pi i,\qquad \int_{C_k''}\frac{d\mu}{\mu-\lambda} = 0.$$

Corollary. The convolutional operator

$$P_k f = \varphi_k * f \tag{37}$$

is a projector of $\mathscr{H}(\bar{D})$ on $\mathscr{E}_{\lambda_k}^{(m_k)}$.

Definition 4. The *Leontiev expansion*, determined by the entire function of exponential type $E(\lambda)$ with zeros λ_k, $k=1, 2, \ldots$, of the locally holomorphic functions on its indicator diagram $\bar{D}$ is said to be the correspondence

$$f(z) \sim \sum_{k=1}^{\infty} f(z) * \varphi_k(z), \tag{38}$$

where φ_k are defined by (35).

If we introduce the Leontiev interpolating function [56], p. 237

$$\omega_E(\mu, f) = \frac{1}{2\pi i} \int_{\Gamma} \gamma(\zeta) d\zeta \int_0^{\zeta} e^{\mu\sigma} f(\zeta - \sigma) d\sigma, \tag{39}$$

we can prove

Lemma 7. *For every $f \in \mathscr{H}(\bar{D})$ and for each positive integer k*

$$P_k f(z) = (f * \varphi_k)(z) = \frac{1}{2\pi i} \int_{C_k} \frac{\omega_E(\mu, f) e^{\mu z} d\mu}{E(\mu)}. \tag{40}$$

Proof. Using (35), we get

$$(f * \varphi_k)(z) = -\frac{1}{2\pi i} \int_{C_k} \frac{f(z) * e^{\mu z}}{E(\mu)} d\mu.$$

But directly from (3) we obtain

$$f(z) * e^{\mu z} = \frac{1}{2\pi i} \int_{\Gamma} \gamma(\zeta) d\zeta \int_{\zeta}^{z} e^{\mu(z+\zeta-\sigma)} f(\sigma) d\sigma$$

$$= \left[\frac{1}{2\pi i} \int_{\Gamma} \gamma(\zeta) e^{\mu\zeta} d\zeta \right] \left[\int_0^z e^{\mu(z-\sigma)} f(\sigma) d\sigma \right]$$

$$- \frac{e^{\mu z}}{2\pi i} \int_{\Gamma} \gamma(\zeta) d\zeta \int_0^{\zeta} e^{\mu\sigma} f(\zeta - \sigma) d\sigma$$

$$= E(\mu) \int_0^z e^{\mu(z-\sigma)} f(\sigma) d\sigma - e^{\mu z} \omega_E(\mu, f).$$

Therefore

$$(f*\varphi_k)(z)=-\int_0^z\left[\frac{1}{2\pi i}\int_{C_k} e^{\mu(z-\sigma)}d\mu\right]f(\sigma)d\sigma+\frac{1}{2\pi i}\int_{C_k}\frac{\omega_E(\mu, f)e^{\mu z}}{E(\mu)}d\mu$$

and the lemma is proved, since $\int_{C_k} e^{\mu(z-\sigma)}d\mu=0$. □

Leontiev [73], p. 241 gives his expansion in the form

$$f(z)\sim\sum_{k=1}^{\infty}\frac{1}{2\pi i}\int_{C_k}\frac{\omega_E(\mu, f)e^{\mu z}}{E(\mu)}d\mu, \tag{41}$$

which, as we have just shown, completely coincides with (38). It is not difficult to perceive that

$$\frac{1}{2\pi i}\int_{C_k}\frac{\omega_E(\mu, f)e^{\mu z}}{E(\mu)}d\mu=\sum_{j=0}^{m_k-1}a_{k,j}z^j e^{\lambda_k z}$$

and that (38) is an expansion of the form

$$f(z)\sim\sum_{k=1}^{\infty}\left(\sum_{j=0}^{m_k-1}a_{k,j}\,z^j\right)e^{\lambda_k z}. \tag{42}$$

To Leontiev [73], p. 255, is due the following *uniqueness theorem,* concerning expansion (38):

If the entire function of exponential type $E(\lambda)$ has infinitely many zeros, and if $f*\varphi_k=0$, for each $k>0$ then $f(z)\equiv 0$.

Definition 5. A multiplier sequence of the formal Leontiev expansion (38) in $\mathcal{H}(\overline{D})$ is said to be every numerical sequence $\{\mu_k\}_{k=1}^{\infty}$, such that for each $f\in\mathcal{H}(\overline{D})$ the series

$$\sum_{k=1}^{\infty}\mu_k(f*\varphi_k)(z)$$

is a formal Leontiev expansion of a function of $\mathcal{H}(\overline{D})$.

Denoting this function by Mf, we have

$$Mf(z)\sim\sum_{k=1}^{\infty}\mu_k(f*\varphi_k)(z). \tag{43}$$

In other words, each multiplier sequence $\{\mu_k\}_{k=1}^{\infty}$ of (38) defines an operator $M:\mathcal{H}(\overline{D})\rightarrow\mathcal{H}(\overline{D})$, such that

$$(Mf)*\varphi_k=\mu_k(f*\varphi_k). \tag{44}$$

Every such operator M shall be named a multiplier of the formal Leontiev expansion (38). Now we aim at characterizing the multipliers of the formal Leontiev expansion (38), and thus the multiplier sequences too.

Theorem 2.2.3. If $M: \mathscr{H}(\bar{D}) \to \mathscr{H}(\bar{D})$ *is a multiplier of the formal Leontiev expansion* (38), *then M is a multiplier of the corresponding convolution* (31) *and hence it admits a representation of the form*

$$Mf = \mu f + m * f \tag{45}$$

with $\mu = \text{const}$ *and with* $m \in \mathscr{H}(\bar{D})$ *with a Leontiev expansion of the form*

$$m(z) \sim \sum_{k=1}^{\infty} \nu_k \varphi_k(z). \tag{46}$$

Conversely, each operator of the form (45) *with a function* $m \in \mathscr{H}(\bar{D})$, *having a Leontiev expansion of the form* (46), *is a multiplier of the formal Leontiev expansion* (38).

Proof. The first part of the theorem states that $M(f*g) = (Mf)*g$ for arbitrary f, $g \in \mathscr{H}(\bar{D})$ with (31) as $*$. Indeed, for an arbitrary integer $k>0$ we have

$$[M(f*g)-(Mf)*g]*\varphi_k = \mu_k(f*g)*\varphi_k$$
$$-[(Mf)*\varphi_k]*g = \mu_k(f*g)*\varphi_k - (\mu_k f)*\varphi_k*g = 0.$$

Now, from the Leontiev uniqueness theorem follows the multiplier relation $M(f*g) = (Mf)*g$. Then, by Theorem 2.1.11, the operator M has the form $Mf = \mu f(z) + (m*f)(z)$ with $\mu = \text{const}$, and with $m = (M\{1\})'$. The k-th term in the Leontiev expansion (38) of m has the form

$$m*\varphi_k = (M\{1\})'*\varphi_k = [M\{1\})*\varphi_k]'$$
$$-\frac{d}{dz}\{(FM\{1\})*\varphi_k\} = M\varphi_k - \mu\varphi_k = (\mu_k - \mu)\varphi_k,$$

where μ is the constant $FM\{1\}$, and $\{\mu_k\}_{k=1}^{\infty}$ is the multiplier sequence which defines M.

The converse is almost evident. □

2.3. A CONVOLUTION FOR THE GENERAL RIGHT INVERSE OF THE BACKWARD SHIFT OPERATOR IN SPACES OF LOCALLY HOLOMORPHIC FUNCTIONS

Along with the differentiation operator, an important role in modern analytic functions theory is played by the *backward shift operator* (see [53])

$$U^* f(z) = \frac{f(z)-f(0)}{z}.$$

It is very much like the differentiation operator. The most general linear right inverse of U^*, considered in a space $\mathscr{H}(\bar{D})$ of locally holomorphic functions on a star-like closed domain $\bar{D}$, has the form

$$Lf(z)=zf(z)+\Phi(f)$$

with an arbitrary linear functional Φ on $\mathcal{H}(\bar{D})$.

A non-trivial convolution $(f*g)(z)$ of L in $\mathcal{H}(\bar{D})$ can be given in a remarkably simple form. This convolution can be used in the same way as convolution (3) from sect. 2.2. Since the considerations here parallel those for the differentiation operator, we give only the basic results.

2.3.1. The linear right inverses of the backward shift operator in a space of locally holomorphic functions. Let D be a finite simple-connected domain in the complex plane, containing the origin $z=0$. Then, by $\mathcal{H}(\bar{D})$ we denote the space of locally holomorphic functions on the closure $\bar{D}$ of D. Each element of $\mathcal{H}(\bar{D})$ is a class of holomorphic functions, every two of them coinciding on a domain containing $\bar{D}$. $\mathcal{H}(\bar{D})$ is a DF-space. The topology in $\mathcal{H}(\bar{D})$ is introduced as the inductive hull topology. Let $\{O_n\}_{n=1}^{\infty}$ be a decreasing sequence of domains with $O_n \supset \bar{O}_{n+1}$ such that every open set containing $\bar{D}$ contains some O_n. By $\mathcal{B}(O_n)$ we denote the space of bounded holomorphic functions on O_n with the supremum norm

$$\|f\|_n = \sup_{z \in O_n} |f(z)|.$$

Then we regard $\mathcal{H}(\bar{D})$ as the inductive limit of the spaces $\mathcal{B}(O_n)$. It is well known that the corresponding inductive topology is such that a sequence of locally holomorphic functions $\tilde{f}_n$ tends to $\tilde{f} \in \mathcal{H}(\bar{D})$ iff there exists an open set $O \supset \bar{D}$ and representatives f_n and f of the classes $\tilde{f}_n$, $\tilde{f}$ which are regular on O, and $f_n \to f$ on O, i. e. uniformly on each compact subset of O. A detailed study of the topology of $\mathcal{H}(\bar{D})$ can be seen in Köthe [69], pp. 375-381.

Now we need a characterization of the continuous linear functionals in $\mathcal{H}(\bar{D})$. A linear map Φ: $\mathcal{H}(\bar{D}) \to \mathbf{C}$ is continuous iff $\tilde{f}_n \to \tilde{0}$ implies $\Phi(\tilde{f}_n) \to 0$. To Sebastaõ-e-Silva [89] is due the following representation theorem for the continuous linear functionals on $\mathcal{H}(\bar{D})$:

A map Φ: $\mathcal{H}(\bar{D}) \to \mathbf{C}$ is a continuous linear functional on $\mathcal{H}(\bar{D})$ iff it admits a representation of the form

$$\Phi(\tilde{f}) = \frac{1}{2\pi i} \int_{\Gamma_f} E(\zeta) f(\zeta) d\zeta \tag{1}$$

with a function $E(\lambda)$, holomorphic on the complement of $\bar{D}$, including in the point $\lambda=\infty$, and Γ_f is a contour, containing $\bar{D}$ in its inside, and lying in the domain of the representative f of $\tilde{f}$.

Since in the sequel we consider continuous functionals and operators only, then usually we drop the adjective "continuous." Moreover, for the sake of simplicity we write $\Phi(f)$ instead of $\Phi(\tilde{f})$.

Now, following the analogy between the backward shift operator U^* and the differentiation operator d/dz, we shall study the right inverse linear operators of U^*. If $\Phi(f)$ is a linear functional on $\mathcal{H}(\bar{D})$, then we can define a right inverse operator L of U^* by means of the simplest boundary value problem

$$U^*u=f, \quad \Phi(u)=0,$$

taking $u=Lf$. This problem has a solution for each $f\in\mathscr{H}(\bar{D})$ iff $\Phi(\{1\})\neq 0$. In this case we may assume $\Phi(\{1\})=1$, without any loss of generality Then it is easy to see that

$$Lf(z)=zf(z)-\Phi_\zeta\{\zeta f(\zeta)\} \tag{2}$$

is the right inverse of U^* we are seeking. It is equally easy to show that each linear right inverse of the backward shift operator U^* has the form (2). We state this as

Theorem 2.3.1. *A linear operator $L: \mathscr{H}(\bar{D})\to\mathscr{H}(\bar{D})$ is a right inverse of the backward shift operator*

$$U^*f(z)=\frac{f(z)-f(0)}{z} \tag{3}$$

in $\mathscr{H}(\bar{D})$ iff it admits a representation of the form (2) *with a linear functional $\Phi: \mathscr{H}(\bar{D})\to \mathrm{C}$, such that $\Phi(\{1\})=1$.*

Proof. It is evident that each linear right inverse operator L of U^* in $\mathscr{H}(\bar{D})$ has the form

$$Lf(z)=zf(z)+\chi(f) \tag{4}$$

with a linear functional χ in $\mathscr{H}(\bar{D})$. The defining projector F of L as a right inverse of (3) (see 1.3.4) is

$$Ff(z)=f(z)-LU^*f(z)=f(0)-\chi(U^*f). \tag{5}$$

It is seen that F is a linear functional on $\mathscr{H}(\bar{D})$ with $F(\{1\})=1$. It is a matter of a simple verification to see that

$$\chi(f)=-F_\zeta\{\zeta f(\zeta)\}, \tag{6}$$

and taking $\Phi=F$, we get (2).

The functional Φ can be expressed by χ without the use of U^* as

$$\Phi(f)=[1-\chi(\{1\})]f(0)+\chi(f). \tag{7}$$

All is clear. □

2.3.2. A class of convolutions in $\mathscr{H}(\bar{D})$, connected with the backward shift operator. The main result in this section is

Theorem 2.3.2. *Let $\Phi: \mathscr{H}(\bar{D})\to\mathrm{C}$ be a non-zero linear continuous functional on $\mathscr{H}(\bar{D})$. Then, the operation*

$$(f*g)(z)=\Phi_\zeta\left\{\frac{[zf(z)-\zeta f(\zeta)]\,[zg(z)-\zeta g(\zeta)]}{z-\zeta}\right\} \tag{8}$$

is a separately continuous, bilinear, commutative and associative operation in $\mathscr{H}(\bar{D})$, such that the convolutional algebra $[\mathscr{H}(\bar{D}), *]$ is annihilators-free.

Proof. The separate continuity of (8) becomes clear, provided we take the contour in representation (1) lying in the intersection of the domains of the representatives f and g of locally holomorphic functions $\tilde{f}$ and $\tilde{g}$, and containing $\bar{D}$ in its inside. The bilinearity and commutativity of (8) are evident too. The associativity of (8) is far from evident. If $\lambda \notin \bar{D}$ is an arbitrary complex number, we denote

$$\varphi_\lambda(z) = \frac{1}{z-\lambda}. \tag{9}$$

Then, if $\lambda \neq \mu$, we get easily

$$(\varphi_\lambda * \varphi_\mu)(z) = \frac{\lambda\mu[E(\lambda)\varphi_\mu(z) - E(\mu)\varphi_\lambda(z)]}{\lambda-\mu}, \tag{10}$$

where $E(\lambda)$ is the Fantappie indicatrix

$$E(\lambda) = \Phi_\zeta\left\{\frac{1}{\zeta-\lambda}\right\} \tag{11}$$

of the functional Φ. Then, for pairwise different $\lambda, \mu, \nu \in \mathbf{C}$ we get easily from (10)

$$(\varphi_\lambda * \varphi_\mu) * \varphi_\nu = \lambda\mu\nu\left\{\frac{E(\mu)E(\nu)}{(\lambda-\mu)(\lambda-\nu)}\varphi_\lambda + \frac{E(\nu)E(\lambda)}{(\mu-\nu)(\mu-\lambda)}\varphi_\mu + \frac{E(\lambda)E(\mu)}{(\nu-\lambda)(\nu-\mu)}\varphi_\nu\right\}.$$

Hence

$$(\varphi_\lambda * \varphi_\mu) * \varphi_\nu = \varphi_\lambda * (\varphi_\mu * \varphi_\nu). \tag{12}$$

Now we shall prove the general associativity relation $(f*g)*h = f*(g*h)$ for arbitrary $f, g, h \in \mathscr{H}(\bar{D})$. Let Γ_f, Γ_g and Γ_h be contours containing $\bar{D}$ inside, lying in the domains of f, g and h correspondingly, and such that they do not intersect. If Γ is that of the contours Γ_f, Γ_g and Γ_h which lies inside the other two, then for a point z inside Γ, we have

$$f(z) = -\frac{1}{2\pi i}\int_{\Gamma_f} f(\lambda)\varphi_\lambda(z)d\lambda, \quad g(z) = -\frac{1}{2\pi i}\int_{\Gamma_g} g(\mu)\varphi_\mu(z)d\mu,$$

$$h(z) = -\frac{1}{2\pi i}\int_{\Gamma_h} h(\nu)\varphi_\nu(z)d\nu,$$

according to the Cauchy integral theorem. Now, let us multiply (12) by $-(2\pi i)^{-3}f(\lambda)g(\mu)h(\nu)$ and integrate on the contours Γ_f, Γ_g and Γ_h with respect to λ, μ and ν correspondingly. We get the associativity relation $(f*g)*h = f*(g*h)$, using the continuity of (8).

Finally we shall show that (8) is annihilators-free. From the assumption that Φ is a non-zero functional it follows the existence of a $\lambda_0 \in \mathbf{C}\setminus\bar{D}$, such that $E(\lambda_0) = \Phi_\zeta\{1/(\zeta-\lambda_0)\} \neq 0$. Indeed, if we assume

$$\Phi_\zeta\left\{\frac{1}{\zeta-\lambda}\right\} = 0 \tag{13}$$

for each $\lambda \in \mathbf{C} \setminus \bar{D}$, then if $f(z)$ is a representative of an arbitrary element of $\mathscr{H}(\bar{D})$, let us multiply (13) by $-f(\lambda)/2\pi i$ and integrate on a contour Γ_f containing $\bar{D}$ inside and lying in the domain of f. Thus we get

$$\Phi_\zeta\left\{\frac{1}{2\pi i}\int_{\Gamma_f}\frac{f(\lambda)d\lambda}{\lambda-z}\right\}=0,$$

i. e. $\Phi(f)=0$ for each $f \in \mathscr{H}(\bar{D})$. Thus, there exists a $\lambda_0 \in \mathbf{C} \setminus \bar{D}$, such that $E(\lambda_0) \neq 0$. Then $\varphi_{\lambda_0}(z)=1/(z-\lambda_0)$ is a non-divisor of zero of (8). Indeed, it is easy to see that the convolutional operator

$$L_0 f = \varphi_{\lambda_0} * f \tag{14}$$

is a right inverse of the operator

$$U_0^* = \frac{1}{E(\lambda_0)}\left[U^* - \frac{1}{\lambda_0} I\right] \tag{15}$$

in $\mathscr{H}(\bar{D})$, where I is the identity operator in $\mathscr{H}(\bar{D})$. Then, if for some $f \in \mathscr{H}(\bar{D})$ we have $\varphi_{\lambda_0} * f = 0$, i. e. $L_0 f = 0$, then applying U_0^* to the last equality, we get $f=0$. Hence $\varphi_{\lambda_0}(z)=1/(z-\lambda_0)$ is a non-zero non-divisor of zero of (8). Therefore, (8) is annihilators-free. □

Now we shall prove some properties of operation (8).

Lemma 1. *If $f, g \in \mathscr{H}(\bar{D})$, then*

$$\Phi(f * g) = 0. \tag{16}$$

Proof. Denote

$$h(\zeta, z) = \frac{[zf(z)-\zeta f(\zeta)]\,[zg(z)-\zeta g(\zeta)]}{z-\zeta}. \tag{17}$$

Since $h(\zeta, z) = -h(z, \zeta)$, then $\Phi(f*g) = \Phi_z\Phi_\zeta\{h(\zeta, z)\} = \Phi_\zeta\Phi_z\{h(\zeta, z)\} = -\Phi_\zeta\Phi_z\{h(z, \zeta)\} = -\Phi(f*g)$ and hence $\Phi(f*g)=0$. □

Lemma 2. *If $\lambda_0 \in \mathbf{C} \setminus \bar{D}$ is such that $E(\lambda_0)=\Phi_\zeta\{1/(\zeta-\lambda_0)\} \neq 0$, then the linear span of the convolutional powers $\{\varphi_{\lambda_0}^{*n}\}_{n=1}^{\infty}$ is dense in $\mathscr{H}(\bar{D})$.*

Proof. In fact, it is asserted that $\varphi_{\lambda_0}(z)=1/(z-\lambda_0)$ is a cyclic element of the convolutional operator $L_0 f = \varphi_{\lambda_0} * f$. Since $0 \in \bar{D}$, then $\lambda_0 \neq 0$. It is easy to see that $\varphi_{\lambda_0}^{*n}$ is a polynomial of $1/(z-\lambda_0)$ exactly of n-th degree, i. e.

$$\varphi_0^{*n}(z) = \sum_{k=1}^{n} \frac{c_{n,k}}{(z-\lambda_0)^k}, \qquad c_{n,n} \neq 0. \tag{18}$$

Now let $f(z)$ be a representative of an arbitrary element of $\mathscr{H}(\bar{D})$. If ψ is the transformation $\psi: z \longrightarrow \frac{1}{z-\lambda_0}$, then the function $g(\zeta)=f(\lambda_0+1/\zeta)$ is an element of $\mathscr{H}(\overline{\psi(D)})$. According to the Runge approximation theorem, there exists a polynomial sequence $p_n(\zeta) = \sum_{k=1}^{n} d_{n,k}\zeta^k$ converging to $g(\zeta)$ in the topology of $\mathscr{H}(\overline{\psi(D)})$. Then the sequence $q_n(z)=p_n(1/(z-\lambda_0))$ converges to $f(z)$ in $\mathscr{H}(\bar{D})$. Hence, the span of $\{\varphi_{\lambda_0}^{*n}\}_{n=1}^{\infty}$ is dense in $\mathscr{H}(\bar{D})$. □

Lemma 3. *If* $f,\ g\in\mathscr{H}(\bar{D})$, *then*

$$U^*(f*g)=(U^*f)*g+\Phi(f)g. \tag{19}$$

Proof. We have

$$U^*(f*g)-(U^*f)*g=\Phi_\zeta\left\{\frac{h(\zeta,z)-h(\zeta,0)}{z}-\frac{[f(z)-f(\zeta)][zg(z)-\zeta g(\zeta)]}{z-\zeta}\right\},$$

where $h(\zeta,z)$ is given by (17). After some elementary algebra we get $f(\zeta)g(z)$ for the expression under the functional sign, thus proving (19). □

Now we are ready to characterize the multipliers of the convolutional algebra $[\mathscr{H}(\bar{D}),*]$.

Theorem 2.3.3. *An operator* $M:\mathscr{H}(\bar{D})\to\mathscr{H}(\bar{D})$ *is a multiplier of convolution* (8), *iff it admits a representation of the form*

$$Mf(z)=\mu f(z)+(m*f)(z) \tag{20}$$

with $\mu=$const, *and with* $m\in\mathscr{H}(\bar{D})$. *Here* $*$ *stands for operation* (8).

Proof. Since $*$ is annihilators-free, then by Theorem 1.3.1 the multipliers of (8) form a commutative ring. Let $M:\ \mathscr{H}(\bar{D})\to\mathscr{H}(\bar{D})$ be an arbitrary multiplier of (8). If $\lambda_0\in\mathbb{C}\setminus\bar{D}$ is chosen so that $E(\lambda_0)=\Phi_\zeta\{1/(\zeta-\lambda_0)\}\neq 0$, then the convolutional operator $L_0f=\varphi_{\lambda_0}*f$ is a multiplier of (8) too. Therefore $ML_0=L_0M$, and hence

$$ML_0f=L_0Mf=(M\varphi_{\lambda_0})*f.$$

But as we have seen in the proof of Theorem 2.3.2 L_0 is a right inverse operator of

$$U_0^*=(U^*-I/\lambda_0)/E(\lambda_0)$$

and hence

$$Mf=U_0^*[(M\varphi_{\lambda_0})*f].$$

From Lemma 3 we get the desired representation (20) with $\mu=\Phi(M\varphi_{\lambda_0})/E(\lambda_0)$ and with $m=U_0^*M\varphi_{\lambda_0}$. □

The converse assertion that each operator of the form (20) is a multiplier of (8) is evident.

Theorem 2.3.4. *Representation* (20) *is uniquely determined by the multiplier* M.

Proof. Indeed, if $Mf=\mu_1f+m_1*f=\mu_2f+m_2*f$ are two representations of a multiplier $M:\ \mathscr{H}(\bar{D})\to\mathscr{H}(\bar{D})$ of (8) in the form (20), then applying Φ to both sides of the last equation, using Lemma 1, we get:

$$\mu_1\Phi(f)=\mu_2\Phi(f).$$

Taking $f\in\mathscr{H}(\bar{D})$ such that $\Phi(f)\neq 0$, we get $M_1=M_2$. Then, $(m_1-m_2)*f=0$ for each $f\in\mathscr{H}(\bar{D})$. But by Theorem 3.3.1 the operation $*$ is annihilators-free and hence $m_1=m_2$. □

2.3.3. The commutant of the backward shift operator in an invariant hyperplane. In 2.1.4 we have found a convolutional representation of the commutant of the differentiation operator in an invariant hyperplane. Here we shall solve the analogous problem for the backward shift operator U^* in $\mathscr{H}(\bar{D})$.

Let $\Phi: \mathscr{H}(\bar{D}) \to \mathbf{C}$ be a non-zero linear functional in $\mathscr{H}(\bar{D})$. Let $*$ be the corresponding convolution (8). By Theorem 2.3.3, each multiplier M of the convolutional algebra $[\mathscr{H}(\bar{D}), *]$ has the form $Mf = \mu f + m * f$. From Lemma 1 it follows that the hyperplane

$$\mathscr{H}_\Phi = \{f \in \mathscr{H}(\bar{D}) : \Phi(f) = 0\} \tag{21}$$

is an invariant subspace of M.

Another observation is that M commutes with U^* in $\mathscr{H}_\Phi$. Indeed, from (20) we get

$$U^*Mf = \mu U^*f + m * U^*f + \Phi(f)m = MU^*f,$$

provided $\Phi(f) = 0$.

Now we shall show that these are the only operators with the invariant subspace $\mathscr{H}_\Phi$, which commute with the backward shift operator in $\mathscr{H}_\Phi$.

Theorem 2.3.5. *Let $\Phi: \mathscr{H}(\bar{D}) \to \mathbf{C}$ be a non-zero linear functional, and $\mathscr{H}_\Phi = \{f \in \mathscr{H}(\bar{D}) : \Phi(f) = 0\}$. Then a linear operator $M: \mathscr{H}(\bar{D}) \to \mathscr{H}(\bar{D})$ with the hyperplane $\mathscr{H}_\Phi$ as an invariant subspace commutes with the backward shift operator U^* in $\mathscr{H}_\Phi$ iff M is a multiplier of convolution* (8), *i. e. iff M has a representation of the form* (20).

Proof. Only the first part of the statement remains to be proved. Let $M: \mathscr{H}(\bar{D}) \to \mathscr{H}(\bar{D})$ be a linear operator, such that $\Phi(Mf) = 0$ when $\Phi(f) = 0$, and $MU^*f = U^*Mf$ for each $f \in \mathscr{H}(\bar{D})$ with $\Phi(f) = 0$. Let us choose a $\lambda_0 \in \mathbf{C} \setminus \bar{D}$ such that $E(\lambda_0) = \Phi_\zeta\{1/(\zeta - \lambda_0)\} \neq 0$. Then we shall show that the requirement that M and U^* commute in $\mathscr{H}_\Phi$ is equivalent to the requirement that M and $L_0 = (\varphi_{\lambda_0})*$ commute in the whole space $\mathscr{H}(\bar{D})$. Indeed, if $f \in \mathscr{H}(\bar{D})$, then let $g = L_0 f$. Since $\Phi(g) = 0$ by Lemma 1, then by assumption $MU_0^*g = U_0^*Mg$, i. e. $Mf = U_0^*(ML_0 f)$. Therefore, $y = ML_0 f$ is the solution of the equation $U_0^* y = Mf$, satisfying the "initial" condition $\Phi(y) = 0$, since by assumption $\mathscr{H}_\Phi$ is an invariant subspace of M. But the only such solution is $y = L_0(Mf)$ and hence $ML_0 f = L_0 Mf$, i. e. M and L_0 commute in $\mathscr{H}(\bar{D})$. The converse is evident.

By Lemma 2, the convolutional operator $L_0 f = \varphi_{\lambda_0} * f$ has φ_{λ_0} as a cyclic element. Then, by Theorem 1.3.11, the commutant of L_0 coincides with the multiplier ring of (8). Hence M is a multiplier of the convolutional algebra $[\mathscr{H}(\bar{D}), *]$.

The theorem is proved. □

2.3.4. Convolutions of the right inverse operators of the backward shift operator in $\mathscr{H}(\bar{D})$. In 2.3.1 we have seen that each linear right inverse operator L of U^* in $\mathscr{H}(\bar{D})$ has a representation of the form

(22) $$Lf(z)=zf(z)-\Phi_\zeta\{\zeta f(\zeta)\}$$

with a linear functional Φ in $\mathscr{H}(\bar{D})$, normalized by the condition $\Phi\{1\}=1$.,

The defining projector $F=I-LU^*$ of L is $Ff=f(z)-LU^*f(z)=\Phi(f)$ i. e. the functional Φ.

Theorem 2.3.6. *The operation* (8) *is a convolution of the right inverse operator* (22) *of* U^* *in* $\mathscr{H}(\bar{D})$, *such that the representation*

(23) $$Lf=\{1\}*f$$

holds.

Proof. Since we have already proved that (8) is a bilinear, commutative and associative operation in $\mathscr{H}(\bar{D})$, in order to assert that it is a convolution of L, it is sufficient to show that (23) holds. This is a matter of a simple verification. □

Now we give one more convolution of L in $\mathscr{H}$.

Theorem 2.3.7. *The operation*

(24) $$(f\tilde{*}g)(z)=U^*(f*g)=\Phi(f)g(z)+\Phi(g)f(z)-\Phi(fg)$$
$$+z\Phi_\zeta\left\{\frac{[f(z)-f(\zeta)][g(z)-g(\zeta)]}{z-\zeta}\right\}$$

is a convolution of L *in* $\mathscr{H}(\bar{D})$, *such that* $\{1\}$ *is the unit of the convolutional algebra* $[\mathscr{H}(\bar{D}),\tilde{*}]$, *i. e.* $\{1\}\tilde{*}f=f$ *for each* $f\in\mathscr{H}(\bar{D})$.

Proof. The bilinearity and commutativity of (24) are evident. To show the associativity only the identity

(25) $$[U^*(f*g)]*h=f*[U^*(g*h)]$$

should be proved. Using (19), we get

$$U^*(f*g*h)=[U^*(f*g)]*h+\Phi(f*g)h=f*[U^*(g*h)]+\Phi(g*h)f.$$

But by Lemma 1, $\Phi(f*g)=\Phi(g*h)=0$, and thus (25) is proved.

The convolution relation $L(f\tilde{*}g)=(Lf)\tilde{*}g$ follows from (25) by the substitution $h=\{1\}$. □

Since the right inverse operator (22) of U^* has the function $\{1\}$ as a cyclic element, then its commutant in $\mathscr{H}(\bar{D})$ coincides with the multiplier ring of (8) or (24). Hence a linear operator $M:\mathscr{H}(\bar{D})\to\mathscr{H}(\bar{D})$ commutes with $Lf(z)=zf(z)-\Phi_\zeta\{\zeta f(\zeta)\}$ with $\Phi\{1\}=1$ iff it admits a representation of the form

(26) $$Mf=\mu f+m*f$$

or

(27) $$Mf=\tilde{m}\tilde{*}f.$$

Now we can characterize all separately continuous convolutions of L in $\mathscr{H}(\bar{D})$.

Theorem 2.3.8. *An operation* $f\hat{*}g$ *in* $\mathscr{H}(\bar{D})$ *is a separately continuous convolution of* (22) *in* $\mathscr{H}(\bar{D})$ *iff it has a representation of the form*

$$f \hat{*} g = r \tilde{*} (f \tilde{*} g) \tag{28}$$

with $r \in \mathscr{H}(\bar{D})$, *where* $\tilde{*}$ *is convolution* (24).

Proof. For a fixed $f \in \mathscr{H}(\bar{D})$, the operator $M_f g = f \hat{*} g$ commutes with L and hence it is a multiplier of $\tilde{*}$. Then from $g = \{1\} \tilde{*} g$ we get $M_f g = (M_f\{1\}) \tilde{*} g = (M_{\{1\}} f) \tilde{*} g = M_{\{1\}}(f \tilde{*} g)$. The operator $M_{\{1\}}$ has the same representation $M_{\{1\}} h = M_{\{1\}}(\{1\} \tilde{*} h)$ and hence

$$f \hat{*} g = (\{1\} \hat{*} \{1\}) \tilde{*} (f \tilde{*} g),$$

thus proving representation (28). The converse is evident. □

2.3.5. Multiplier projectors on spectral subspaces of a right inverse of the backward shift operator. The general consideration from 1.3.2 can be transferred to the right inverses of the backward shift operator U^* in $\mathscr{H}(\bar{D})$. Let

$$Lf(z) = zf(z) - \Phi_\zeta\{\zeta f(\zeta)\} \tag{29}$$

be a right inverse of U^* in $\mathscr{H}(\bar{D})$ with a linear functional Φ, satisfying the condition $\Phi\{1\} = 1$.

In the previous subsection we have seen that the operation

$$(f * g)(z) = \Phi_\zeta\left\{\frac{[zf(z) - \zeta f(\zeta)][zg(z) - \zeta g(\zeta)]}{z - \zeta}\right\} \tag{30}$$

is a convolution of L in $\mathscr{H}(\bar{D})$, such that $Lf = \{1\} * f$.

If we are interested in characterizing the divisors of zero of (30), we should first find the eigenelements of (29). It is easy to see that a complex number λ is an eigenvalue of L iff it is a zero of the *Fantappie indicatrix*

$$E(\lambda) = \Phi_\zeta\left\{\frac{1}{\zeta - \lambda}\right\}, \quad \lambda \in \mathbb{C} \setminus \bar{D} \tag{31}$$

of Φ. The spectral multiplicity of each eigenvalue of L is finite and it is equal to the multiplicity of λ as a zero of $E(\lambda)$.

Let m_k be the multiplicity of an eigenvalue λ_k of L. The function

$$\varphi_{\lambda_k}(z) = \frac{1}{z - \lambda_k} \tag{32}$$

is the corresponding eigenelement of L. The spectral subspace $\mathscr{H}_{\lambda_k}$ of L, corresponding to the eigenvalue λ_k, is the kernel-space

$$\mathscr{H}_{\lambda_k} = \ker (L - \lambda_k I)^{m_k}. \tag{33}$$

It is easy to see that $\mathscr{H}_{\lambda_k}$ is the linear span of the functions

$$\frac{1}{(z - \lambda_k)^j}, \quad j = 1, 2, \dots, m_k.$$

Theorem 1.3.5 ensures the existence of an element $u_k \in \mathscr{H}_{\lambda_k}$, such that he convolutional operator $P_k f = u_k * f$ is a projector of $\mathscr{H}(\bar{D})$ onto $\mathscr{H}_{\lambda_k}$.

But it is desirable to have this multiplier projector in explicit form. The following theorem gives this element in an explicit form, resembling the corresponding result (35) from 2.2.3 for the differentiation operator.

Theorem 2.3.9. *Let $\lambda_k \in \mathbf{C}\setminus\bar{D}$ be a zero of the Fantappie indicatrix (31) of a linear functional Φ in $\mathscr{H}(\bar{D})$ with $\Phi\{1\}=1$. If Γ_k is a contour in $\mathbf{C}\setminus\bar{D}$, containing λ_k, and none of the other zeros of $E(\lambda)$, then the function*

$$u_k(z)=\frac{1}{2\pi i}\int_{\Gamma_k}\frac{d\lambda}{\lambda^2E(\lambda)(z-\lambda)} \tag{34}$$

satisfies the idempotential relation

$$u_k^{*2}=u_k. \tag{35}$$

Proof. Let, as in the proof of Lemma 6 of 2.2.3, Γ_k' and Γ_k'' be two contours in $\mathbf{C}\setminus\bar{D}$, such that Γ_k'' is contained in Γ_k', and such that each of them contains only the zero λ_k of $E(\lambda)$, and none of the other zeros of $E(\lambda)$. Then, writing

$$u_k(z)=\frac{1}{2\pi i}\int_{\Gamma_k'}\frac{d\lambda}{\lambda^2E(\lambda)(z-\lambda)},\qquad u_k(z)=\frac{1}{2\pi i}\int_{\Gamma_k''}\frac{d\mu}{\mu^2E(\mu)(z-\mu)},$$

we get

$$u_k^{*2}(z)=\frac{1}{(2\pi i)^2}\iint_{\Gamma_k'\times\Gamma_k''}\frac{(\varphi_\lambda*\varphi_\mu)(z)}{\lambda^2\mu^2E(\lambda)E(\mu)}\,d\lambda d\mu,$$

where $\varphi_\lambda(z)=1/(z-\lambda)$ and $\varphi_\mu(z)=1/(z-\mu)$. Formula (10) gives

$$(\varphi_\lambda*\varphi_\mu)(z)=\frac{\lambda\mu[E(\lambda)\varphi_\mu(z)-E(\mu)\varphi_\lambda(z)]}{\lambda-\mu}.$$

Thus we get

$$u_k^{*2}(z)=\frac{1}{(2\pi i)^2}\int_{\Gamma_k''}\frac{\varphi_\mu(z)}{\mu E(\mu)}\,d\mu\int_{\Gamma_k'}\frac{d\lambda}{\lambda(\lambda-\mu)}-\frac{1}{(2\pi i)^2}\int_{\Gamma_k'}\frac{\varphi_\lambda(z)d\lambda}{\lambda E(\lambda)}\int_{\Gamma_k''}\frac{d\mu}{\mu(\lambda-\mu)}$$

$$=\frac{1}{2\pi i}\int_{\Gamma_k''}\frac{\varphi_\mu(z)}{\mu^2E(\mu)}\,d\mu=u_k(z)$$

since $\int_{\Gamma_k'}d\lambda/\lambda(\lambda-\mu)=2\pi i/\mu$ and $\int_{\Gamma_k''}d\mu/\mu(\lambda-\mu)=0$.

The theorem is proved. □

Corollary. The operator $P_kf=u_k*f$ is a multiplier projector of $\mathscr{H}(\bar{D})$ on the kernel space $\mathscr{H}_{\lambda_k}=\ker(L-\lambda_kI)^{m_k}$. In particular, if λ_k is a simple zero of the Fantappie indicatrix $E(\lambda)$, then

$$P_k f = -\frac{1}{\lambda_k^2 E'(\lambda_k)}\left\{\frac{1}{z-\lambda_k}\right\} * f = \left[\frac{1}{2\pi i E'(\lambda_k)\lambda_k}\int_{\Gamma_k}\frac{\lambda E(\lambda) f(\lambda)}{\lambda-\lambda_k}\,d\lambda\right]\frac{1}{z-\lambda_k}. \tag{36}$$

The first assertion follows from (35). Formula (36) can easily be obtained using (34) and the explicit form (8) of the convolution $*$.

Open problem. If $\lambda_1, \lambda_2, \ldots$ are all the zeros of $E(\lambda)$ in $\mathbf{C}\setminus\bar{D}$, find necessary and sufficient conditions for the totality of the multiplier projectors system $P_k f = u_k * f$, $k=1, 2, \ldots$.

Most of the results in this section are contained in Dimovski and Mineff [49].

2.4. CONVOLUTIONS AND COMMUTANTS OF THE GELFOND-LEONTIEV INTEGRATION OPERATOR AND OF ITS INTEGER POWERS

To Gelfond and Leontiev [58] is due the following generalization of the differentiation operator $D=d/dz$ for analytic functions $f(z)=\sum_{n=0}^{\infty} a_n z^n$ in a disc $|z|<R$:

$$D_\varrho f(z) = \sum_{n=1}^{\infty} a_n \frac{\Gamma(n/\varrho+1)}{\Gamma((n-1)/\varrho+1)} z^{n-1}$$

with a fixed positive constant ϱ. The *Gelfond-Leontiev integration operator* is said to be the right inverse operator l_ϱ of D_ϱ, defined by

$$l_\varrho f(z) = \sum_{n=0}^{\infty} a_n \frac{\Gamma(n/\varrho+1)}{\Gamma((n+1)/\varrho+1)} z^{n+1}.$$

In this section we extend the above definition of the Gelfond-Leontiev integration operator for spaces of analytic functions in domains star-like with respect to the origin, and find a convolution of l_ϱ in such a space. By means of this convolution the commutant of l_ϱ is found. Finally, a convolutional representation of the commutant of any fixed integer power l_ϱ^m of l_ϱ is found. Thus some results for the integration operator $l=l_1$, due to Raichinov [86] and [88], are generalized.

2.4.1. An integral representation of the Gelfond-Leontiev integration operator in a space of analytic functions in a domain star-like with respect to the origin. Let $\mathscr{H}(\Delta_R)$ be the space of the functions $f(z)$, which are analytic in the disc $\Delta_R=\{z: |z|<R\}$, and let ϱ be a fixed positive number. In Gelfond and Leontiev [58] the following generalization of the differentiation operation is introduced.

Definition 1. Let $\varrho>0$. Then the Gelfond-Leontiev differentiation operator in $\mathscr{H}(\Delta_R)$ is said to be the map

$$D_\varrho\left[\sum_{n=0}^{\infty} a_n z^n\right] = \sum_{n=1}^{\infty} a_n \frac{\Gamma(n/\varrho+1)}{\Gamma((n-1)/\varrho+1)} z^{n-1}. \tag{1}$$

The operator D_ϱ is right invertible, and each of its linear right inverses L_ϱ has the form

(2) $$L_\varrho f(z)=l_\varrho f(z)+\Phi(f),$$

where l_ϱ is the initial right inverse of D_ϱ, defined by

(3) $$l_\varrho\left[\sum_{n=0}^{\infty} a_n z^n\right]=\sum_{n=0}^{\infty} a_n \frac{\Gamma(n/\varrho+1)}{\Gamma((n+1)/\varrho+1)} z^{n+1},$$

and Φ is a linear functional on $\mathscr{H}(\Delta_R)$.

Definition 2. The linear operator l_ϱ: $\mathscr{H}(\Delta_R)\to\mathscr{H}(\Delta_R)$, defined by (3), is said to be the Gelfond-Leontiev integration operator in $\mathscr{H}(\Delta_R)$.

Now we give an integral representation of l_ϱ, which allows l_ϱ to be extended to more general spaces of analytic functions.

Lemma 1. *In $\mathscr{H}(\Delta_R)$ the representation formula*

(4) $$l_\varrho f(z)=\frac{z}{\Gamma(1/\varrho)}\int_0^1 (1-\sigma)^{1/\varrho-1} f(z\sigma^{1/\varrho})d\sigma$$

holds.

Proof. It is sufficient to verify (4) for an arbitrary power z^n, $n=0$, 1, 2,.... We have

$$l_\varrho\{z^n\}=\frac{z^{n+1}}{\Gamma(1/\varrho)}\int_0^1 (1-\sigma)^{1/\varrho-1}\sigma^{n/\varrho}d\sigma=\frac{\Gamma(n/\varrho+1)}{\Gamma((n+1)/\varrho+1)}z^{n+1}.$$

Then, if $f(z)=\sum_{n=0}^{\infty} a_n z^n$ is an arbitrary function of $\mathscr{H}(\Delta_R)$, we get $l_\varrho f(z)$ $=\sum_{n=0}^{\infty} a_n l_\varrho\{z^n\}=\sum_{n=0}^{\infty} a_n \frac{\Gamma(n/\varrho+1)}{\Gamma((n+1)/\varrho+1)} z^{n+1}$, and the lemma is proved. □

By representation (4) l_ϱ can be defined not only in $\mathscr{H}(\Delta_R)$ but in a space $\mathscr{H}(\Delta)$ of the analytic function in a domain Δ star-like with respect to the origin 0.

Definition 3. The integral operator l_ϱ, defined by (4) in the space $\mathscr{H}(\Delta)$ of the analytic functions $f(z)$ in a star-like domain Δ, is said to be the Gelfond-Leontiev integration operator in $\mathscr{H}(\Delta)$.

In particular, if $\varrho=1$, we get from (4)

$$l_1 f(z)=z\int_0^1 f(z\sigma)\,d\sigma=\int_0^z f(\zeta)d\zeta,$$

i. e. l_1 is the usual Volterra integration operator in $\mathscr{H}(\Delta)$.

Remark. By (4) a linear operator in the space $\mathscr{C}(\Delta)$ of the continuous functions on a real interval Δ, containing the point 0, is also defined.

It is useful to have such an extension of the Gelfond-Leontiev differentiation operator D_ϱ, given by (1) for $\varrho>0$. This can in fact be done for all $\varrho>0$, but the result has a simpler form for $\varrho\geqq 1$. That is why we restrict ourselves to the case $\varrho \geqq 1$, though most of the following considerations are valid without this restriction.

Lemma 2. *The integro-differential operator*

$$(5)\quad D_\varrho f(z) = \begin{cases} \dfrac{1}{\varrho}\left(z\dfrac{d}{dz}+\varrho\right) z^{-1}\displaystyle\int_0^1 \dfrac{(1-\sigma)^{-1/\varrho}}{\Gamma(1-1/\varrho)}\,[\,f(z\sigma^{1/\varrho}) - f(0)]\,d\sigma, & \varrho>1, \\ \dfrac{d}{dz} f(z), \quad \varrho=1, \end{cases}$$

defined on $\mathscr{H}(\Delta)$ with star-like Δ, coincides with the Gelfond-Leontiev differentiation operator D_ϱ, defined by (1) *for $\Delta=\Delta_R$*

Proof. It is sufficient to show this for a non-negative integer power z^n. According to (5), we have

$$D_\varrho\{z^n\} = \frac{(n-1+\varrho)}{\varrho\Gamma(1-1/\varrho)}\, z^{n-1}\int_0^1 (1-\sigma)^{1/\varrho}\,\sigma^{n/\varrho}\,d\sigma = \frac{\Gamma(n/\varrho+1)}{\Gamma((n-1)/\varrho+1)}\, z^{n-1},$$

i. e. in $\mathscr{H}(\Delta_R)$ the operator (5) coincides with (1). □

That is why for $\varrho\geqq 1$ we will call the operator D_ϱ, defined by (5), the Gelfond-Leontiev differentiation operator. In this case l_ϱ is a right inverse of D_ϱ in $\mathscr{H}(\Delta)$.

The defining projector $Ff=f-l_\varrho D_\varrho f$ of l_ϱ as a right inverse of (5) is

$$(6)\qquad Ff(z)=f(0).$$

2.4.2. A convolution of the Gelfond-Leontiev integration operator in $\mathscr{H}(\Delta)$. Next we show that l_ϱ has a convolution in $\mathscr{H}(\Delta)$.

Theorem 2.4.1. *The operation*

$$(7)\qquad (f*g)(z) = \frac{1}{\varrho}\left(z\frac{d}{dz}+\varrho\right)\int_0^1 f\,[z(1-\tau)^{1/\varrho}]\,g(z\tau^{1/\varrho})\,d\tau$$

is a convolution of the Gelfond-Leontiev integration operator l_ϱ in $\mathscr{H}(\Delta)$, in which the constant function $\{1\}$ is the unit, i. e.

$$(8)\qquad \{1\}*f=f.$$

Proof. The bilinearity and commutativity of (7) are evident. Now we shall show that (7) is an associative operation in $\mathscr{H}(\Delta)$. To this end, we verify first the associativity relation $(f*g)*h=f*(g*h)$ for non-negative integer powers $f(z)=z^m$, $g(z)=z^n$ and $h(z)=z^p$. Since

$$(9)\qquad \{z^j\}*\{z^k\} = \frac{\Gamma(j/\varrho+1)\Gamma(k/\varrho+1)}{\Gamma[(j+k)/\varrho+1]}\, z^{j+k},$$

then

$$(\{z^m\}*\{z^n\})*\{z^p\} = \frac{\Gamma(m/\varrho+1)\Gamma(n/\varrho+1)\Gamma(p/\varrho+1)}{\Gamma[(m+n+p)/\varrho+1]}\, z^{m+n+p}.$$

From the symmetry of this expression, it follows

$$(10)\qquad (\{z^m\}*\{z^n\})*\{z^p\} = \{z^m\}*(\{z^n\}*\{z^p\})$$

for arbitrary integers $m, n, p=0, 1, 2, \dots$. Then, due to the bilinearity of (7), the associativity relation is satisfied by polynomials. But from the Runge approximation theorem and from the evident continuity of (7), it follows that the associativity relation holds in the whole space $\mathscr{H}(\Delta)$.

It remains only to be shown that (7) is in fact a convolution of l_ϱ in $\mathscr{H}(\Delta)$. To this end we should establish the representation formula

$$l_\varrho f=\left\{\frac{\varrho z}{\Gamma(1/\varrho)}\right\}*f. \tag{11}$$

It can be seen either directly, or by approximation, first verifying it for non-negative integer powers z^n. □

Corollary. Convolution (7) of l_ϱ has a representation of the form

$$(f*g)(z)=f(0)g(z)+\frac{z}{\varrho}\int_0^1 f'[z(1-\tau)^{1/\varrho}]g(z\tau^{1/\varrho})(1-\tau)^{1/\varrho-1}d\tau. \tag{12}$$

The identity (12) can be shown either by polynomial approximation verifying it first for powers, or directly, using the representation

$$\int_0^1 f[z(1-\tau)^{1/\varrho}]g(z\tau^{1/\varrho})d\tau=\varrho z^{-\varrho}\int_0^z f[(z^\varrho-\zeta^\varrho)^{1/\varrho}]g(\zeta)\zeta^{\varrho-1}d\zeta,$$

where the integration is on the segment connecting the points $\zeta=0$ and $\zeta=z$. Then

$$(f*g)(z)=z^{1-\varrho}\frac{d}{dz}\int_0^z f[(z^\varrho-\zeta^\varrho)^{1/\varrho}]g(\zeta)\zeta^{\varrho-1}d\zeta$$

$$=f(0)g(z)+z^{1-\varrho}\int_0^z f'[(z^\varrho-\zeta^\varrho)^{1/\varrho}]z^{\varrho-1}(z^\varrho-\zeta^\varrho)^{1/\varrho-1}g(\zeta)\zeta^{\varrho-1}d\zeta,$$

which is (12).

2.4.3. The commutant of the Gelfond-Leontiev integration operator in $\mathscr{H}(\Delta)$. Now, following the general scheme of 1.3.5, we can find a convolutional representation of the commutant of the Gelfond-Leontiev integration operator. The representation is a generalization of Raichinov's representation (see [86])

$$Mf(z)=m(0)f(z)+\int_0^z m'(z-\zeta)f(\zeta)d\zeta$$

of the commutant of the Volterra integration operator in $\mathscr{H}(\Delta)$.

Let $M: \mathscr{H}(\Delta)\to\mathscr{H}(\Delta)$ be an arbitrary continuous linear operator, which commutes with l_ϱ in $\mathscr{H}(\Delta)$, i. e. such that

$$Ml_\varrho=l_\varrho M. \tag{13}$$

Then, by Theorem 1.3.11, M is a multiplier of convolution (7), since l_ϱ is an operator with a cyclic element in $\mathscr{H}(\Delta)$. Indeed, the constant function

$\{1\}$ is a cyclic element of l_ϱ, since the span of $\{l_\varrho^n\{1\}\}_{n=0}^{\infty}$ coincides with the space of the polynomials in $\mathscr{H}(\Delta)$. But, according to Runge's approximation theorem, the polynomials are dense in $\mathscr{H}(\Delta)$.

Theorem 2.4.2. *A linear operator $M: \mathscr{H}(\Delta)\to\mathscr{H}(\Delta)$ commutes with the Gelfond-Leontiev integration operator l_ϱ in $\mathscr{H}(\Delta)$, defined by* (4), *iff it admits a representation of the form*

$$Mf(z)=(m*f)(z) \tag{14}$$

with $m(z)=M\{1\}\in\mathscr{H}(\Delta)$, where by $*$ operation (7) *is denoted.*

Proof. Let $M: \mathscr{H}(\Delta)\to\mathscr{H}(\Delta)$ be a linear operator, commuting with l_ϱ in $\mathscr{H}(\Delta)$. Then, by Theorem 1.3.11, M is a multiplier of (7). But from $f=\{1\}*f$ we get at once

$$Mf=(M\{1\})*f,$$

thus proving representation (14).

Conversely, each convolutional operator of the form (14) obviously commutes with l_ϱ. □

Corollary. A linear operator $M: \mathscr{H}(\Delta)\to\mathscr{H}(\Delta)$ commutes with the Gelfond-Leontiev integration operator l_ϱ in $\mathscr{H}(\Delta)$ iff it admits a representation of the form

$$Mf(z)=\mu f(z)+z\int_0^1 m[z(1-\tau)^{1/\varrho}]f[z\tau^{1/\varrho}](1-\tau)^{1/\varrho-1}d\tau \tag{15}$$

with $\mu=\text{const}$, and with $m(z)\in\mathscr{H}(\Delta)$.

According to the corollary of the previous theorem, (15) is only (14), developed by (12).

Both (14) and (15) are simpler representations than the representation found by Tkachenko [102] for the commutant of the Gelfond-Leontiev integration operator.

2.4.4. A convolutional representation of the commutant of a fixed integer power of the Gelfond-Leontiev integration operator. Let m be a fixed positive integer. If $m>1$, then the operators $M: \mathscr{H}(\Delta)\to\mathscr{H}(\Delta)$ commuting with l_ϱ^m in $\mathscr{H}(\Delta)$ are much more than those commuting with l_ϱ. We have already seen this for the square l^2 of the Volterra integration operator l (see example 2 of 1.3.5). Raichinov [88] has found an explicit representation of the commutant of l^m in a m-symmetric domain Δ. A domain Δ is said to be m-symmetric, iff $z\in\Delta$ implies $\omega z\in\Delta$, where ω is an arbitrary m-th root of the unity, i. e. a complex number with $\omega^m=1$.

We shall find a representation of the commutant of l_ϱ^m in an m-symmetric domain too.

Definition 3. Let ω be a primitive m-th root of unity. If $k=0, 1, 2, \dots, m-1$, then $\mathscr{H}_k(\Delta)$ is the subspace of $\mathscr{H}(\Delta)$, consisting of all functions $f(z)\in\mathscr{H}(\Delta)$, such that

$$f(\omega z)=\omega^k f(z) \quad \text{for} \quad z\in\Delta. \tag{16}$$

Lemma 3. *The space $\mathscr{H}(\Delta)$ of the analytic functions in a m-symmetric and star-like domain with respect to the origin, is a direct sum of the subspaces $\mathscr{H}_k(\Delta)$, $k=0, 1, 2,\dots, m-1$, i. e.*

$$\mathscr{H}(\Delta)=\mathscr{H}_0(\Delta)\oplus\mathscr{H}_1(\Delta)\oplus\cdots\oplus\mathscr{H}_{m-1}(\Delta). \tag{17}$$

Proof. If $f(z)\in\mathscr{H}(\Delta)$ is arbitrary, then we denote

$$f_k(z)=\frac{1}{m}\sum_{j=0}^{m-1}\omega^{-kj}f(\omega^j z),\quad k=0,\ 1,\dots,\ m-1. \tag{18}$$

We shall show that $f_k(z)\in\mathscr{H}_k(\Delta)$, and that

$$f(z)=\sum_{k=0}^{m-1}f_k(z). \tag{19}$$

Indeed,

$$f_k(\omega z)=\frac{1}{m}\sum_{j=0}^{m-1}\omega^{-kj}f(\omega^{j+1}z)=\frac{1}{m}\sum_{j'=1}^{m}\omega^{-k(j'-1)}f(\omega^{j'}z)=\omega^k f_k(z),$$

since $\omega^{-mk}=1$. Moreover,

$$\sum_{k=0}^{m-1}f_k(z)=\sum_{k=0}^{m-1}\frac{1}{m}\sum_{j=0}^{m-1}\omega^{-kj}f(\omega^j z)=\frac{1}{m}\sum_{j=0}^{m-1}\left(\sum_{k=0}^{m-1}\omega^{-kj}\right)f(\omega^j z)=f(z),$$

since

$$\sum_{k=0}^{m-1}\omega^{-kj}=\begin{cases}0 & \text{for}\quad j=1,2,\dots,m-1,\\ m & \text{for}\quad j=0.\end{cases}$$

Let us remark that (16) does not depend on the special choice of the primitive m-th root of unity ω. Indeed, if ζ with $\zeta^m=1$ is another m-th primitive root of unity, then $\zeta=\omega^q$ with an integer exponent q, $(q,\ m)=1$. Then

$$f_k(\zeta z)=f_k(\omega^q z)=\omega^{kq}f_k(z)=\zeta^k f_k(z).$$

Only the uniqueness of (19) remains to be shown. Let

$$f(z)=\sum_{k=0}^{m-1}g_k(z)$$

be a representation of $f(z)\in\mathscr{H}(\Delta)$ with some $g_k(z)\in\mathscr{H}_k(\Delta)$, $k=0,\ 1,\ 2,\dots,\ m-1$. Then

$$\sum_{j=0}^{m-1}\omega^{-pj}f(\omega^j z)=\sum_{j=0}^{m-1}\omega^{-pj}\sum_{k=0}^{m-1}g_k(\omega^j z)$$

$$=\sum_{j=0}^{m-1}\sum_{k=0}^{m-1}\omega^{-pj}\omega^{jk}g_k(z)=\sum_{k=0}^{m-1}\left(\sum_{j=0}^{m-1}\omega^{j(k-p)}\right)g_k(z)=mg_p(z),$$

since

$$\sum_{j=0}^{m-1}\omega^{j(k-p)}=\begin{cases}0 & \text{for}\quad k\neq p,\\ m & \text{for}\quad k=p.\end{cases}$$

Thus the lemma is proved. □

Each of the spaces $\mathscr{H}_k(\Delta)$, $k=0, 1, 2, \ldots, m-1$ is an invariant subspace for l_ϱ^m.

Lemma 4. *The operations*

$$f \overset{k}{*} g = D_\varrho^k(f * g), \quad k=0, 1, 2, \ldots, m-1, \tag{20}$$

where D_ϱ is the Gelfond-Leontiev differentiation operator in $\mathscr{H}(\Delta)$, given by (5), *are convolutions of l_ϱ^m in its invariant subspaces $\mathscr{H}_k(\Delta)$, correspondingly. Then, if $f \in \mathscr{H}_k(\Delta)$, the identities*

$$\left\{\frac{z^k}{\Gamma(k/\varrho+1)}\right\} \overset{k}{*} f = f \tag{21}$$

hold. In other words, $e_k(z)=\dfrac{z^k}{\Gamma(k/\varrho+1)}$ is the unit of the convolution $\overset{k}{*}$ in $\mathscr{H}_k(\Delta)$.

Proof. From the fact that $*$ is a convolution of l_ϱ in $\mathscr{H}(\Delta)$ it follows only that $\overset{k}{*}$, $k=0, 1, 2, \ldots, m-1$, are bilinear and commutative operations, but the associativity should be proved separately. It is easy to find that if p, q are non-negative integers, then

$$\{z^{mp+k}\} \overset{k}{*} \{z^{mq+k}\} = \frac{\Gamma[(mp+k)/\varrho+1]\,\Gamma[(mq+k)/\varrho+1]}{\Gamma[(m+p+q)/\varrho+1]} z^{m(p+q)+k}. \tag{22}$$

Now, using (22) we can easily verify the associativity relation $(f \overset{k}{*} g) \overset{k}{*} h) = f \overset{k}{*} (g \overset{k}{*} h)$ for powers of the form $f(z)=z^{mp+k}$, $g(z)=z^{mq+k}$ and $h(z)=z^{mr+k}$. It can easily be seen that the polynomials in $\mathscr{H}_k(\Delta)$ are of the form $P(z^m)z^k$, where $P(\zeta)$ is a polynomial, and that they are dense in $\mathscr{H}_k(\Delta)$.

The identities (21) can be verified either directly, or by means of polynomial approximation. The convolutional relation $l_\varrho^m(f \overset{k}{*} g) = (l_\varrho^m f) \overset{k}{*} g$ follows from the representation

$$l_\varrho^m f = (l_\varrho e_k) \overset{k}{*} f, \tag{23}$$

which can be verified in the same way. □

Lemma 5. *The function $e_k(z)=\dfrac{z^k}{\Gamma(k/\varrho+1)}$ is a cyclic element of l_ϱ^m* in *$\mathscr{H}_k(\Delta)$, $k=0, 1, 2, \ldots, m-1$.*

Proof. Since

$$(l_\varrho^m)^n e_k(z) = C_{k,n} z^{mn+k} \text{ with } C_{k,n} \neq 0,$$

then the linear span of $\{(l_\varrho^m)^n e_k(z)\}_{n=0}^{\infty}$ coincides with the set of the polynomials in $\mathscr{H}_k(\Delta)$. Let $f(z)$ be an arbitrary function from $\mathscr{H}_k(\Delta)$. Then, by Lemma 3, $f(z) = \dfrac{1}{m}\sum_{j=0}^{m-1} \omega^{-kj} f(\omega^j z)$.

If $\{Q_n(z)\}_{n=0}^{\infty}$ is a polynomial sequence, converging to $f(z)$ uniformly on each compact subset of Δ, which can be found according to Runge's theorem, then the polynomial sequence

$$Q_{n,k}(z)=\frac{1}{m}\sum_{j=0}^{m-1}\omega^{-kj}Q_n(\omega^j z),\quad n=0,\,1,\,2,\ldots,$$

consisting of polynomials from $\mathscr{H}_k(\Delta)$, shall converge to $f(z)$ uniformly on each compact subset of Δ too. Hence $e_k(z)$ is a cyclic element of l_ϱ^m in $\mathscr{H}_k(\Delta)$. □

Lemma 6. *Let $M\colon \mathscr{H}(\Delta)\to\mathscr{H}(\Delta)$ be a linear operator commuting with l_ϱ^m in $\mathscr{H}_k(\Delta)$. Then for each $f\in\mathscr{H}_k(\Delta)$ the identity*

$$Mf=(Me_k)\overset{k}{*}f,\quad k=0,\,1,\ldots,m-1, \tag{24}$$

holds,

Proof. From the assumption $Ml_\varrho^m=l_\varrho^m M$ on $\mathscr{H}_k(\Delta)$ in particular it follows that $Ml_\varrho^m e_k=l_\varrho^m Me_k$. From expression (20) of $f\overset{k}{*}g$ it follows that for arbitrary non-negative integers p, q

$$(Ml_\varrho^{mp}e_k)\overset{k}{*}(l_\varrho^{mq}e_k)=(l_\varrho^{mp}e_k)\overset{k}{*}(Ml_\varrho^{mq}e_k). \tag{25}$$

Therefore the multiplier relation $(Mf)\overset{k}{*}g=f\overset{k}{*}(Mg)$ holds for functions of the form $f(z)=\sum_{p=0}^{r}a_p z^{mp+k}$ and $g(z)=\sum_{q=0}^{s}b_q z^{mq+k}$, i. e. for polynomials from $\mathscr{H}_k(\Delta)$. By Lemma 4, the relation

$$(Mf)\overset{k}{*}g=f\overset{k}{*}(Mg) \tag{26}$$

holds for arbitrary $f,\,g\in\mathscr{H}_k(\Delta)$. In particular, if we take $g=e_k$, we get at once (24). □

Theorem 2.5.3. *A linear operator $M\colon \mathscr{H}(\Delta)\to\mathscr{H}(\Delta)$ in the space $\mathscr{H}(\Delta)$ of the analytic functions in an m-symmetric and star-like domain Δ commutes with the m-th power l_ϱ^m of the Gelfond-Leontiev integration operator l_ϱ iff it admits a representation of the form*

$$Mf=\sum_{k=0}^{m-1}D_\varrho^k(m_k * f_k) \tag{27}$$

with arbitrary $m_k=Me_k\in\mathscr{H}(\Delta)$, and with

$$f_k(z)=\frac{1}{m}\sum_{j=0}^{m-1}\omega^{-kj}f(\omega^j z),\quad k=0,\,1,\ldots,\,m-1, \tag{28}$$

where ω is a primitive m-th root of unity.

Proof. By Lemma 3, $f(z)=\sum_{k=0}^{m-1} f_k(z)$ and $Mf=\sum_{k=0}^{m-1} Mf_k$. By Lemma 6, we have $Mf_k=m_k \overset{k}{*} f_k=D_\varrho^k(m_k * f)$, and M has the form (27). It remains to show that each operator of the form (27) commutes with l_ϱ^m in $\mathscr{H}(\Delta)$. Indeed, on the one hand, we have

$$Ml_\varrho^m f=\sum_{k=0}^{m-1} D_\varrho^k(m_k * l_\varrho^m f_k)=\sum_{k=0}^{m-1} l_\varrho^{m-k}(m_k * f_k),$$

and on the other hand,

$$l_\varrho^m Mf=\sum_{k=0}^{m-1} l_\varrho^{m-k}(l_\varrho^k D_\varrho^k)(m_k * f_k)$$

$$=\sum_{k=0}^{m-1} l_\varrho^{m-k}(m_k * l_\varrho^k D_\varrho^k f_k)=\sum_{k=0}^{m-1} l_\varrho^{m-k}(m_k * f_k),$$

since for arbitrary $m_k \in \mathscr{H}(\Delta)$, we have $D_\varrho^k(m_k * f_k)=m_k *(D_\varrho^k f_k)$, provided $f_k \in \mathscr{H}_k(\Delta)$, and $l_\varrho^k D_\varrho^k f_k=D_\varrho^k l_\varrho^k f_k=f_k$.

The theorem is proved. □

Corollary (Raichinov [88]). A linear operator M in $\mathscr{H}(\Delta)$ commutes with the m-th power l^m of the Volterra integration operator l in $\mathscr{H}(\Delta)$ iff it admits a representation of the form

$$Mf(z)=\sum_{k=0}^{m-1} \frac{d^{k+1}}{dz^{k+1}} \int_0^z m_k(z-\zeta) f_k(\zeta) d\zeta \tag{29}$$

with arbitrary $m_k(z) \in \mathscr{H}(\Delta)$, $k=0, 1, 2, \ldots, m-1$, and with $f_k(z)$ given by (28).

2.5. OPERATION CALCULI FOR THE BERNOULLI INTEGRATION OPERATOR

In the previous sections we used the convolutions for integration operators in problems of analysis. Here our aim is to show that these convolutions, like the Duhamel convolution, can be used for developing new operational calculi, some of which may be of practical importance. As a typical example we have taken the Bernoulli integration operator and its convolution. The corresponding operational calculus can be developed to such an extent as the classical Heaviside-Mikusiński operational calculus. A new feature is the abundance of divisors of zero of the corresponding convolution, which causes no troubles. In treating "abstract differential equations", i. e. ordinary linear differential equations with constant coefficients, for the corresponding Taylor boundary value problem a new element is the possibility to treat them in a purely algebraic way in the resonance case as well. This is a specialization of the general scheme of 1.4.5. In particular periodic solutions of ordinary linear differential equations with constant coefficients and of systems of such equations in all feasible cases, can be found effectively.

2.5.1. The ring of the convolutional quotients of the Bernoulli convolution algebra. Here we consider the right inverse operator of the differentiation, defined in $\mathscr{C}([0, 1])$ by the solution of the elementary boundary value problem

(1) $$y'=f, \quad \int_0^1 y(\tau)d\tau=0.$$

This integration operator

(2) $$Lf(t)=\int_0^t f(\tau)d\tau - \int_0^1 (1-\tau)f(\tau)d\tau$$

is named here the *Bernoulli integration operator*, since it is connected with the Bernoulli polynomials $B_n(t)$:

(3) $$L^n\{1\}=\frac{1}{n!}B_n(t), \quad n=0, 1, 2, \ldots .$$

The defining projector of L is the linear functional

(4) $$Ff(t)=f(t)-Lf'(t)=\int_0^1 f(\tau)d\tau.$$

This is a linear functional of the type $Ff=\chi(lf)$ with $\chi(f)=f(1)$. Then using the results of sect. 2.1, one can at once write down a convolution of L in $\mathscr{C}([0, 1])$. This is the operation

(5) $$(f*g)(t)=f(t)\int_0^1 g(\tau)d\tau+g(t)\int_0^1 f(\tau)d\tau$$
$$-\int_0^t f(t-\tau)g(\tau)d\tau-\int_t^1 f(1+t-\tau)g(\tau)d\tau.$$

It is a convolution with a unit, i. e. for all $f\in\mathscr{C}([0, 1])$

(6) $$\{1\}*f=f.$$

The Bernoulli integration operator can be represented as the convolutional operator

(7) $$Lf=\left\{t-\frac{1}{2}\right\}*f.$$

Next we shall characterize the divisors of zero of (5). As we know from 1.3.3, each eigenfunction of L is a divisor of zero of its convolution (5).

Lemma 1. *The eigenfunctions of the Bernoulli integration operator are*

(8) $$\varphi_n(t)=e^{2\pi int},\quad n=\pm1,\ \pm2,\dots,$$

with the simple eigenvalues $\lambda_n=1/2\pi in$ *correspondingly.*

Proof. It is easy to see that if $L\varphi=\lambda\varphi$ with $\varphi\neq0$, then $\lambda\neq0$, i. e. $\lambda=0$ is not an eigenvalue of L. Then φ is a solution of the problem $\varphi'=\varphi/\lambda$, $\int_0^1\varphi(\tau)d\tau=0$. We find easily $\varphi(t)=Ce^{t/\lambda}$ with $e^{1/\lambda}-1=0$. Therefore $\lambda=1/2\pi in$, $n=\pm1,\ \pm2,\ \pm3,\dots$, are the only eigenvalues of L, and they all are simple. □

Lemma 2. *For each* $f\in\mathscr{C}([0,1])$ *the identities*

(9) $$f*\{e^{2\pi int}\}=\chi_n(f)\{e^{2\pi int}\},\quad n=\pm1,\ \pm2,\dots,$$

with

(10) $$\chi_n(f)=\int_0^1(1-e^{-2\pi in\tau})f(\tau)d\tau,$$

hold.

Proof. By a direct check. □

Lemma 3. *The linear functionals* $\chi_n(f)$, *defined by* (10), *are multiplicative with respect to convolution* (5), *i. e.*

(11) $$\chi_n(f*g)=\chi_n(f)\chi_n(g),\quad n=\pm1,\ \pm2,\dots.$$

Proof. Using (9) and the associativity of (5), we get

$$\chi_n(f*g)\{e^{2\pi int}\}=(f*g)*\{e^{2\pi int}\}=f*(g*\{e^{2\pi int}\})$$
$$=f*\{\chi_n(g)e^{2\pi int}\}=(f*\{e^{2\pi int}\})\chi_n(g)=\chi_n(f)\chi_n(g)\{e^{2\pi int}\}.$$

Hence, the functionals χ_n are multiplicative. □

Lemma 4. *The linear functionals* $\{\chi_n\}$, $n=\pm1,\ \pm2,\dots$, *form a total system of linear functionals, i. e. such that* $\chi_n(f)=0$ *for* $f\in\mathscr{C}([0,1])$ *and* $n=\pm1,\ \pm2,\dots$, implies $f(t)\equiv0$.

Proof. Let us denote $g(t)=Lf(t)$. If $\chi_n(f)=0$, then $\chi_n(g)=\chi_n(\{t-1/2\}*f)$ $=\chi_n(\{t-1/2\})\chi_n(f)=0$ for $n=\pm1,\ \pm2,\dots$. Since $\int_0^1 g(\tau)d\tau=0$, by (1), then $\int_0^1 g(\tau)e^{-2\pi in\tau}d\tau=0$ for $n=0,\ \pm1,\ \pm2,\dots$. From a well-known theorem from the theory of the Fourier series it follows that $g(t)\equiv0$. Hence $f(t)=\frac{d}{dt}g(t)\equiv0$. □

Lemma 5. *A function* $f(t)\in\mathscr{C}([0,1])$ *is non-trivial non-divisor of zero of* (5) *iff*

(12) $$\int_0^1(1-e^{-2\pi in\tau})f(\tau)d\tau\neq0\quad\text{for}\quad n=\pm1,\ \pm2,\dots.$$

Proof. Let us assume first that (12) are fulfilled. Using Lemma 3, (12) and $f*g=0$, we get $\chi_n(g)=0$, $n=\pm 1, \pm 2, \ldots$. From Lemma 4 it follows that $g\equiv 0$. The converse easily follows from the multiplicativity of the functionals χ_n. □

Open problem. Describe all divisors of zero of (5) in the space $\mathscr{C}(\Delta)$ with an interval Δ, containing the segment $[0, 1]$ in its inside.

Now let us construct the ring of the convolution quotients of (5) for $\mathscr{C}=\mathscr{C}([0, 1])$. By $\mathscr{C}_*$ we denote the multiplicative set of the non-trivial non-divisors of zero of (5). Since (5) is a convolution with the constant function $\{1\}$ as unit, each multiplier of $*$ is a convolutional operator. Indeed, if M is a multiplier of (5), then from $f=\{1\}*f$ it follows that $Mf=(M\{1\})*f$, and so $M=(M\{1\})*$. Therefore, for the Bernoulli convolution algebra no multiplier "problem" exists at all and it is a matter of taste whether we consider the convolution quotients ring of (5) or the multiplier quotients ring. We use the same notation M for them both. But for definiteness let us speak in terms of convolution quotients.

The Bernoulli integration operator should be identified in the convolution quotients ring $\mathsf{M}=\mathscr{C}_*^{-1}\mathscr{C}$ with the function $\{t-1/2\}$, i. e. $L=\{t-1/2\}$.

A basic role in the corresponding operational calculus is played by the inverse element

$$S=\frac{1}{L} \tag{13}$$

of L in M. If for the sake of brevity we denote $f_0=\int_0^1 f(\tau)d\tau$, then the connection between f' and Sf is given by the formula

$$f'=Sf-f_0S. \tag{14}$$

Indeed, for $f\in\mathscr{C}^1([0, 1])$, (14) is equivalent to

$$Lf'=f-f_0.$$

Relation (14) can be generalized to

$$f^{(n)}=S^nf-\sum_{k=0}^{n-1} f_0^{(k)}S^{n-k}.$$

But $f_0^{(k)}=\int_0^1 f^{(k)}(\tau)d\tau=f^{(k-1)}(1)-f^{(k-1)}(0)$, and hence

$$f^{(n)}=S^nf-S^n\int_0^1 f(\tau)d\tau-\sum_{k=1}^{n-1}[f^{(k-1)}(1)-f^{(k-1)}(0)]S^{n-k}. \tag{15}$$

For the applications of the operational calculus for the Bernoulli integration operator it is important to characterize the divisors of zero in M of the form $P(S)$, where $P(\lambda)$ is a polynomial.

Lemma 6. *If $P(\lambda)=a_0\lambda^n+a_1\lambda^{n-1}+\cdots+a_{n-1}\lambda+a_n$, $a_0\neq 0$, is a polynomial with complex coefficients, then the element $P(S)$ is non-divisor of zero iff $P(2\pi im)\neq 0$ for each integer $m\neq 0$.*

Proof. First we show that if $P(2\pi im)=0$ for some integer $m\neq 0$, then $P(S)$ is a divisor of zero in M. Indeed, then $P(S)=(S-2\pi im)Q(S)$ with a polynomial Q. Further, $P(S)\{e^{2\pi imt}\}=0$, since $(S-2\pi im)\{e^{2\pi imt}\}=0$. This follows from $L\{e^{2\pi imt}\}=\frac{1}{2\pi im}e^{2\pi imt}$. Hence $P(S)$ is a divisor of zero in M.

Conversely, let $P(S)$ be a divisor of zero in M, and let $\lambda_1, \lambda_2, \ldots, \lambda_k$ be the different zeros of $P(\lambda)$. Then at least one of the factors $S-\lambda_m$ of $P(S)$ should be a divisor of zero in M. Let $S-\lambda_m$ be a divisor of zero. Then there exists a function $f\in\mathscr{C}([0, 1])$, $f\neq 0$, such that $(S-\lambda_m)f=0$. But multiplying the last equation by L, we get $f-\lambda_m Lf=0$. This means that f is an eigenfunction of L with eigenvalue λ_m^{-1}. From Lemma 1 it follows that $\lambda_m=2\pi ik_m$ with a certain integer $k_m\neq 0$. The lemma is proved. □

Theorem 2.5.1. *If $\lambda\neq 2\pi in$ for $n=\pm 1, \pm 2, \ldots$, then*

$$\frac{1}{(S-\lambda)^k}=\left\{\frac{(-1)^k}{\lambda^k}+\frac{1}{(k-1)!}\frac{\partial^{k-1}}{\partial\lambda^{k-1}}\left(\frac{e^{\lambda t}}{e^{\lambda}-1}\right)\right\} \tag{16}$$

for each integer $k>0$. In particular,

$$\frac{1}{S-\lambda}=\left\{-\frac{1}{\lambda}+\frac{e^{\lambda t}}{e^{\lambda}-1}\right\}. \tag{17}$$

Proof. Let us first prove (17). Denote the right-hand side of (17) by y. It is easy to verify that y satisfies the equation

$$y(t)-\lambda Ly(t)=t-1/2. \tag{18}$$

But $L=\{t-1/2\}$, and we can write (18) as $y-\lambda Ly=L$, thus proving (17)

Let us prove (16). It can be obtained from (17) by k-times formal differentiation with respect to λ. But in order to justify this, we shall use induction on k. Denoting the right-hand side of (16) by $y_k(\lambda, t)$, we obviously have

$$\frac{\partial y_k}{\partial\lambda}=ky_{k+1}. \tag{19}$$

Assuming that (16) is true for some k, we can write it in the form $(1-\lambda L)^k y_k=L^k$, or equivalently,

$$[1-\lambda(t-1/2)]^{*k}\{y_k(\lambda, t)\}=\frac{1}{k!}B_k(t),$$

where the convolutional power is on the variable t. Differentiating the last equation with respect to the parameter λ, we get

$$-k\left(t-\frac{1}{2}\right)*\left\{1-\lambda\left(t-\frac{1}{2}\right)\right\}^{*(k-1)}*y_k+\left\{1-\lambda\left(t-\frac{1}{2}\right)\right\}^{*k}*\frac{\partial y_k}{\partial\lambda}=0.$$

Here we can cancel the non-zero non-divisor of zero $\{1-\lambda(t-1/2)\}^{*(k-1)}$ and get

$$-k\cdot\left\{\left(t-\frac{1}{2}\right)\right\}*y_k+\left\{1-\lambda\left(t-\frac{1}{2}\right)\right\}*\frac{\partial y_k}{\partial\lambda}=0,$$

i. e.

$$-k\,Ly_k+(1-\lambda L)\frac{\partial y_k}{\partial\lambda}=0.$$

Writing this in the form $(\partial y_k/\partial\lambda)=[k/(S-\lambda)]y_k$, from (19) we get

$$y_{k+1}=\frac{1}{S-\lambda}y_k=\frac{1}{(S-\lambda)^{k+1}},$$

thus proving (16) with $k+1$ instead of k. □

Corollary. If $\lambda\neq 2\pi n$, $n=\pm1, \pm2,\ldots$, then

$$(20)\qquad \frac{1}{S^2+\lambda^2}=\left\{\frac{1}{\lambda^2}-\frac{\cos\lambda\left(t-\frac{1}{2}\right)}{2\lambda\sin\frac{\lambda}{2}}\right\}.$$

Representation (20) can be obtained directly from (17), using the partial fractions decomposition

$$\frac{1}{S^2+\lambda^2}=\frac{1}{2\lambda i}\left\{\frac{1}{S-\lambda i}-\frac{1}{S+\lambda i}\right\}.$$

Formulas (15) and (16) permit the Heaviside algorithm for solving initial value problems for ordinary linear differential equations with constant coefficients to be transferred to boundary value problems of the form

$$(21)\qquad P\left(\frac{d}{dt}\right)x=f,\quad x_0^{(k)}=\gamma_k,\ k=0, 1,\ldots, n-1,$$

with a polynomial $P(\lambda)=\alpha_0\lambda^n+\alpha_1\lambda^{n-1}+\cdots+\alpha_{n-1}\lambda+\alpha_n$, $\alpha_0\neq0$, such that $P(2\pi ik)\neq0$ for all integers $k\neq0$, and with given numerical values of γ_k. In this case we have

$$(22)\qquad x_0=\int_0^1 x(\tau)d\tau,\quad x_0^{(k)}=x^{(k-1)}(1)-x^{(k-1)}(0),\ k=1, 2,\ldots, n-1.$$

In the ring M the problem (21) can be reduced to the single equation

$$(23)\qquad P(S)x=f+Q(S)\quad\text{with}\quad Q(\lambda)=\sum_{j=1}^{n}\left(\sum_{k=j}^{n}\alpha_{n-k}\gamma_{k-j}\right)S^j.$$

The degree of Q does not exceed n. Since, by Lemma 6, $P(S)$ is a non-divisor of zero, then the solution of (23) in M is uniquely determined and

$$(24)\qquad x=\frac{1}{P(S)}f+\frac{Q(S)}{P(S)}.$$

Now we shall show that (24) can be interpreted as a function from $\mathscr{C}^n([0, 1])$, which is a solution of the boundary value problem (21). At first we show that the function $\eta=\frac{1}{P(S)}f$ is the solution of (21) with zero boundary value conditions. To this end we use the fact that the function $\varphi(\lambda)=\{\varphi(\lambda, t)\}=\left\{-\frac{1}{\lambda}+\frac{e^{\lambda t}}{e^{\lambda}-1}\right\}$, which represents the partial fraction $1/(S-\lambda)$, satisfies the condition $\varphi_0(\lambda)=\int_0^1 \varphi(\lambda, \tau)d\tau=0$. Further we have the evident property $(f*g)_0 = f_0 g_0$ of convolution (5). If λ_1 is one of the zeros of $P(\lambda)$, then $\eta=\varphi(\lambda_1)*g$ with $g\in\mathscr{C}([0, 1])$. Then $\eta_0=\varphi_0(\lambda_1)g_0=0$. Let us assume that it is already proved that $\eta_0^{(j)}=0$ for $j=0, 1,\ldots, k-1$ with $k<n$. Then from (15) we get

$$\eta^{(k)}=S^k\eta=\frac{S^k}{P(S)}f.$$

The fractional expression $\frac{S^k}{P(S)}$ can be decomposed into a sum of partial fractions of the form $\frac{A_{jk}}{(S-\lambda_j)^k}$, i. e. $\eta^{(k)} = \sum_{j=1}^{m}\sum_{k=1}^{k_j}\frac{A_{jk}}{(S-\lambda_j)^k}f = \sum_{j=1}^{m}\varphi(\lambda_j)*g_j$. Therefore $\eta_0^{(k)}=\sum_{j=1}^{m}\varphi_0(\lambda_j)(g_j)_0$. Hence $\eta_0^{(k)}=0$, thus proving that $\eta_0^{(j)}=0$ for $j=0, 1, 2,\ldots, n-1$.

It remains to prove that $y=Q(S)/P(S)$ is a solution of (21) with $f=0$. Since each of the functions representing a partial fraction is infinitely differentiable, then the corresponding boundary value problem is equivalent to the single equation $P(S)y=Q(S)$. Hence x, determined by (24), is a true solution of (21).

C o r o l l a r y. Under the hypothesis of the theorem, the solution of (21) with zero boundary value conditions $\gamma_k=0$, $k=0, 1,\ldots, n-1$, can be represented in the form

(25) $$x=g*f,$$

where g is the solution of the same problem but with $f=\{1\}$.

The boundary value problem

(26) $$P\left(\frac{d}{dt}\right)x=f,\quad x^{(k)}(1)-x^{(k)}(1)=\gamma_k,\quad k=0, 1,\ldots, n-1,$$

can be reduced to problem (21), provided the polynomial $P(\lambda)=\alpha_0\lambda^n+\alpha_1\lambda^{n-1}+\ldots+\alpha_n$ is such that $P(2\pi ik)\neq 0$ for each integer k. Let us point out that here we assume additionally that $P(0)\neq 0$.

It is not difficult to see that (26) is equivalent to (21) with $\gamma_0 = -\frac{1}{\alpha_n}(\alpha_0\gamma_{n-1}+\alpha_1\gamma_{n-2}+\ldots+\alpha_{n-1}\gamma_1)$. Indeed, if y is a solution of (26), then from the equation $\alpha_0 y^{(n)}+\alpha_1 y^{(n-1)}+\ldots+\alpha_n y=0$ we get $\alpha_0(y^{(n)})_0+\alpha_1(y^{(n-1)})_0+\ldots+\alpha_{n-1}(y')_0+\alpha_n y_0=0$, i. e. $\alpha_0\gamma_{n-1}+\alpha_1\gamma_{n-1}+\ldots+\alpha_{n-1}\gamma_1+\alpha_n\gamma_0=0$. Therefore, y is a solution of (21) with the above γ_0. The converse is clear.

Finally, we shall give one model of the quotient ring M of the Bernoulli convolution algebra $\mathscr{C}([0, 1])$. An analogous model, but for the periodic convolution, is proposed by B o e h m e and W y g a n t [8]. In our case to each $f\in\mathscr{C}([0, 1])$ we associate the two-sided sequence

(27) $$a=(\ldots, a_{-2}, a_{-1}, a_1, a_2, a_3,\ldots)$$

with $a_k=\chi_k(f)=\int\limits_0^1(1-e^{-2\pi ikt})f(t)dt,\ k=\pm1,\ \pm2,\dots$. By Lemma 4 this correspondence is one-to-one. Let us denote it by χ. The image $\chi(\mathscr{C})$ of the Bernoulli convolution algebra $\mathscr{C}=\mathscr{C}([0,1])$ is an algebra of sequences with the usual addition, multiplication by a constant and term-by-term multiplication of sequences.

As we have seen (Lemma 5), a function $f\in\mathscr{C}([0,1])$ is a non-trivial non-divisor of zero iff $\chi_k(f)\neq0$ for all k. Hence the image $\chi(\mathscr{C}_*)$ of the multiplicative set of the non-trivial non-divisors of zero of (5) consists of sequences (27), with only non-zero terms.

Let us consider the ring of sequences of the form (27) with the term-by-term operations

$$(a+b)_k=a_k+b_k,\quad (ab)_k=a_kb_k.$$

In this ring $\mathscr{R}$ it is always possible to divide by all b with $b_n\neq0$ for every n. Then the operation

$$\frac{a}{b}=\left(\dots,\frac{a_{-2}}{b_{-2}},\ \frac{a_{-1}}{b_{-1}},\ \frac{a_1}{b_1},\ \frac{a_2}{b_2},\dots\right) \tag{28}$$

is defined in $\mathscr{R}$ for all a and for b with $b_n\neq0$ for all n.

Theorem 2.5.2. *If* M *is the convolution quotients ring for the Bernoulli convolution* (5), *and if f/g is an arbitrary element of* M, *then the correspondence $\mathscr{T}:\mathrm{M}\to\mathscr{R}$, defined by*

$$\mathscr{T}\left(\frac{f}{g}\right)=\frac{\chi(f)}{\chi(g)}, \tag{29}$$

is a ring isomorphism.

Proof. First we should be aware that (29) is well defined. Indeed, if $f/g=f_1/g_1$, then $f*g_1=g*f_1$ and $\chi_k(f)\chi_k(g_1)=\chi_k(g)\chi_k(f_1)$. Hence $\chi_k(f)/\chi_k(g)=\chi_k(f_1)/\chi_k(g_1)$, i. e. correspondence (29) is well defined. Next we show that it is one-to-one. Let $\mathscr{T}(f/g)=\mathscr{T}(f_1/g_1)$. Then

$$\frac{\chi_k(f)}{\chi_k(g)}=\frac{\chi_k(f_1)}{\chi_k(g_1)}\quad\text{or}\quad \chi_k(f)\chi_k(g_1)-\chi_k(g)\chi_k(f_1)=0$$

for every integer $k\neq0$. The last relation can be written in the form $\chi_k(f*g_1-g*f_1)=0$. Therefore, $f*g_1=g*f_1$ and hence $f/g=f_1/g_1$.

Let us now prove that $\mathscr{T}$ is a map of M on $\mathscr{R}$. If $a\in\mathscr{R}$ is arbitrarily chosen, then for each $k\neq0$ a representation

$$a_k=\frac{b_k}{c_k},\quad\text{with}\quad |b_k|<1/k^2,\quad |c_k|<1/k^2, \tag{30}$$

can be found. Then the functions $f(t)=\sum\limits_{k\neq0}b_ke^{2\pi ikt}$ and $g(t)=\sum\limits_{k\neq0}c_ke^{2\pi ikt}$ as sums of uniformly convergent series are functions from $\mathscr{C}([0,1])$. Then

$$\mathscr{T}(f/g)=\left(\dots,\frac{b_{-2}}{c_{-2}},\ \frac{b_{-1}}{c_{-1}},\ \frac{b_1}{c_1},\ \frac{b_2}{c_2},\dots\right)=a.$$

Thus we have proved the existence of an element $f/g \in \mathrm{M}$, such that it is an inverse image of $a \in \mathscr{R}$.

It is evident that $\mathscr{T}$ is a ring isomorphism.

Obviously, the map $\mathscr{T}$ is an extension of the map χ, since if $f \in \mathscr{C}$, then $f=f/\{1\}$, and

$$[\mathscr{T}(f/\{1\})]_k=\chi_k(f)/\chi_k(1)=\chi_k(f)$$

due to $\chi_k(1)=1$. □

This sequential model of the convolution quotients ring of the Bernoulli convolution (5) can be used for a solution of the analogon of the Mikusiński problem, stated at the end of 1. 2. 3.

Let the real variable λ vary in the segment $[a, b]$. As in 1.2.3, the derivative of a function $f:[a, b]\to \mathrm{M}$ of the form $f(\lambda)=\{\varphi(t, \lambda)\}/\{\psi(t, \lambda)\}$, as convolution quotient, where $\varphi(t, \lambda)$ and $\psi(t, \lambda)$ are defined and continuous on the rectangle $a\leqq\lambda\leqq b$, $0\leqq t\leqq 1$, and have continuous partial derivatives $\varphi_\lambda(t, \lambda)$ and $\psi_\lambda(t, \lambda)$ on λ, is natural to be defined as

$$f'(\lambda)=\frac{\{\varphi_\lambda'(t, \lambda)\}*\{\psi(t, \lambda)\}-\{\varphi(t, \lambda)\}*\{\psi_\lambda'(t, \lambda)\}}{\{\psi(t, \lambda)\}^{*2}}. \tag{31}$$

Theorem 2.5.3. *If $f'(\lambda)=0$ for $a\leqq\lambda\leqq b$, then $f(\lambda)=\mathrm{const}$.*

Proof. The equality $f'(\lambda)=0$ is equivalent to

$$\{\varphi_\lambda(t, \lambda)\}*\{\psi(t, \lambda)\}-\{\varphi(t, \lambda)\}*\{\psi_\lambda(t, \lambda)\}=0, \tag{32}$$

where $*$ denotes the Bernoulli convolution (5). Let us denote for the sake of brevity $\varphi_k(\lambda)=\chi_k\{\varphi(t, \lambda)\}$, and $\psi_k(\lambda)=\chi_k\{\psi_\lambda(t, \lambda)\}$. Then we have $\chi_k\{\varphi_\lambda'\}=\varphi_k'(\lambda)$ and $\chi_k\{\psi_\lambda'\}=\psi_k'(\lambda)$. If k is an arbitrary integer, then from (32) we get

$$\varphi_k'(\lambda)\psi_k(\lambda)-\varphi_k(\lambda)\psi_k'(\lambda)=0, \qquad k=\pm 1,\ \pm 2, \ldots . \tag{33}$$

Since, by assumption, $\psi_k(\lambda)\neq 0$ for each $k\neq 0$, then

$$\frac{\varphi_k(\lambda)}{\psi_k(\lambda)}=\mathrm{const}=\frac{\varphi_k(a)}{\psi_k(a)}, \qquad k=\pm 1,\ \pm 2, \ldots . \tag{34}$$

All these relations are equivalent to the single relation

$$\{\varphi(t, \lambda)\}*\{\psi(t, a)\}=\{\psi(t, \lambda)\}*\{\varphi(t, a)\}.$$

Therefore,

$$f(\lambda)=\frac{\{\varphi(t, \lambda)\}}{\{\psi(t, \lambda)\}}=\frac{\{\varphi(t, a)\}}{\{\psi(t, a)\}}=f(a)$$

and hence f is a constant.

Thus all is proved. □

2.5.2. Rings of multiplier quotients of subalgebras of the Bernoulli convolution algebra. The algorithm developed in the last section is inapplicable to the resonance case, when $P(2\pi i k_l)=0$ for some integers k_l, $l=1, 2, \ldots, m$. Nevertheless, it can be so modified as to embrace the resonance case too. The idea of this modification is due to Grozdev [63].

Let $k\neq 0$ be a given integer. Denote by $\mathscr{C}_k$ the subspace of $\mathscr{C}([0, 1])$, consisting of those functions f, for which

$$\chi_k(f)=\int_0^1(1-e^{-2\pi ikt})f(t)dt=0,$$

i. e.

(35) $$\mathscr{C}_k=\{f:f\in\mathscr{C}([0,\ 1]),\ \ \chi_k(f)=0\}.$$

If $k_1, k_2, \ldots, k_m$ are different integers, $k_i\neq 0$, then for the sake of brevity we introduce the multi-subscript $\mathbf{k}=(k_1, k_2, \ldots, k_m)$. We use the notation

(36) $$\mathscr{C}_{\mathbf{k}}=\mathscr{C}_{k_1}\cap\mathscr{C}_{k_2}\cap\ldots\cap\mathscr{C}_{k_m}.$$

In other words $f\in\mathscr{C}_{\mathbf{k}}$, iff $f\in\mathscr{C}([0,\ 1])$ and $\chi_{k_l}(f)=0$ for $l=1, 2, \ldots, m$.

Lemma 7. *Each $\mathscr{C}_k$ is an ideal in the Bernoulli convolution algebra $\mathscr{C}([0,\ 1])$.*

Proof. Due to (36) it is enough to prove that $\mathscr{C}_k$ for an integer $k\neq 0$ is an ideal. This is almost evident, since if $f\in\mathscr{C}([0,\ 1])$, and $g\in\mathscr{C}_k$, then

$$\chi_k(f*g)=\chi_k(f)\chi_k(g)=0$$

and hence $f*g\in\mathscr{C}_k$. □

Lemma 8. *If $\mathbf{k}=(k_1, k_2,\ldots, k_m)$ is a multi-subscript with non-zero mutually different integers, then the function*

(37) $$e_{\mathbf{k}}=\{1-e^{2\pi ik_1t}-\ldots-e^{2\pi ik_mt}\}$$

is the unit of the Bernoulli convolution in $\mathscr{C}_{\mathbf{k}}$.

Proof. It is easy to verify that $e_{\mathbf{k}}\in\mathscr{C}_{\mathbf{k}}$. Hence

$$e_{\mathbf{k}}*f=\{1\}*f-\sum_{l=1}^{m}\chi_{k_l}(f)\{e^{2\pi ik_lt}\}=f.\ \square$$

Corollary. The Bernoulli convolution algebra $\mathscr{C}_{\mathbf{k}}$ is annihilators-free.

Lemma 9. *The operator $P_{\mathbf{k}}: \mathscr{C}\to\mathscr{C}_{\mathbf{k}}$, defined by*

(38) $$P_{\mathbf{k}}f=f-\sum_{l=1}^{m}\chi_{k_l}(f)\{e^{2\pi ik_lt}\},$$

is a projector commuting with the Bernoulli convolution (5).

Proof. Let us prove the projector relation $P_{\mathbf{k}}^2=P_{\mathbf{k}}$. We have $P_{\mathbf{k}}^2f$ $=P_{\mathbf{k}}(P_{\mathbf{k}}f)=P_{\mathbf{k}}f-\sum_{l=1}^{m}\chi_{k_l}(P_{\mathbf{k}}f)P_{\mathbf{k}}\{e^{2\pi ik_lt}\}=P_{\mathbf{k}}f$, since $\chi_{k_l}(P_{\mathbf{k}}f)=0$. From the representation

(39) $$P_{\mathbf{k}}f=e_{\mathbf{k}}*f$$

it follows that $P_{\mathbf{k}}$ is a multiplier of the Bernoulli convolution (5), i. e. it commutes with it. □

Lemma 10. *If P: $\mathscr{C}\to\mathscr{C}_{\mathbf{k}}$ is a projector, commuting with the Bernoulli convolution in $\mathscr{C}$, then $P=P_{\mathbf{k}}$.*

Proof. Since P is a multiplier of the Bernoulli convolution (5), it admits a representation of the form $Pf=p*f$ with $p=P\{1\}$. Since for each $f\in\mathscr{C}_k$ we should have $Pf=f$, then $Pe_k=e_k$. But $Pe_k=p*e_k$ and hence $p=e_k$. Then $Pf=e_k*f=P_kf$. □

Lemma 11. *The restriction of an arbitrary multiplier $M:\mathscr{C}\to\mathscr{C}$ of the Bernoulli convolution algebra $\mathscr{C}([0,1])$ to $\mathscr{C}_k$ is an operator in $\mathscr{C}_k$.*

Proof. Since each multiplier of the Bernoulli convolution (5) has the form $Mf=m*f$ with $m\in\mathscr{C}$, then the assertion follows from Lemma 7. □

The restriction of such a multiplier M to $\mathscr{C}_k$ will be denoted by M_k. In particular, L_k is the restriction of the Bernoulli integration operator L to $\mathscr{C}_k$. In this case we have

$$L_kf=r_k*f$$

with

$$r_k(t)=t-\frac{1}{2}-\sum_{l=1}^{m}\frac{1}{2\pi ik_l}e^{2\pi ik_lt}. \tag{40}$$

Lemma 12. *The restriction M_k of an arbitrary multiplier $Mf=q*f$ of the Bernoulli convolution algebra $\mathscr{C}$ to $\mathscr{C}_k$ has a representation of the form*

$$M_kf=q_k*f \tag{41}$$

with $q_k=P_kq=q-\sum_{l=1}^{m}\chi_{k_l}(q)\{e^{2\pi ik_lt}\}$.

Proof. Since $M_kf=P_kMf$, then $M_kf=(P_kq)*f$ and the assertion follows from (38). □

Now let us consider a Bernoulli convolution algebra $\mathscr{C}_k$. It is the projection of the Bernoulli convolution algebra $\mathscr{C}=\mathscr{C}([0,1])$ by (38), i. e. $\mathscr{C}_k=P_k(\mathscr{C})$.

Let M_k be the multiplier quotients ring of $\mathscr{C}_k$ with respect to the non-trivial non-divisors of zero in the multiplier ring $\mathfrak{M}_k$. In a sense, $\mathfrak{M}_k$ is identical with $\mathscr{C}_k$, which justifies treatment of the elements of $\mathscr{C}_k$ as elements of $\mathfrak{M}_k$. So, writing, e. g. e_k, we shall understand the multiplier operator $\{e_k\}*$, i. e. the identity in $\mathfrak{M}_k$. The identity of M_k shall be e_k too. When our considerations are restricted only in M_k, it is convenient to write 1 instead of e_k. If $f,g\in\mathscr{C}_k$, then considering f and g as multipliers of the Bernoulli convolution algebra, we shall write fg instead of $(f*g)*$. But before proceeding further, let us characterize the non-trivial non-divisors of zero in $\mathscr{C}_k$ and in M_k too.

Lemma 13. *A function $f\in\mathscr{C}_k$, $f\neq0$, is a non-divisor of zero of the Bernoulli convolution* (5) *in $\mathscr{C}_k$ iff $\chi_n(f)\neq0$ for each integer $n\neq0$, with $n\neq k_l$ $(l=1,2,\ldots,m)$.*

Proof. As the proof of Lemma 5. □

Corollary. The element r_k, given by (40), is not a divisor of zero in $\mathscr{C}_k$. Indeed, if $n\neq k_l$, $l=1,\ldots,m$, then $\chi_n(r_k)=\frac{1}{2\pi in}\neq0$.

It follows in particular that the restriction L_k of the Bernoulli integration operator L to $\mathscr{C}_k$, is not a divisor of zero in M_k. Hence it has an inverse element in M_k:

(42) $$S_{\mathbf{k}}=\frac{1}{L_{\mathbf{k}}}.$$

Theorem 2.5.4. *Let $\mathsf{M}_{\mathbf{k}}$ be the multiplier quotients ring of the Bernoulli convolution subalgebra $\mathscr{C}_{\mathbf{k}}$ with $\mathbf{k}=(k_1, k_2, \ldots, k_m)$. Then the elements $S_{\mathbf{k}}-2\pi ik_l$, $l=1, 2, \ldots, m$, are non-divisors of zero in $\mathsf{M}_{\mathbf{k}}$, and*

(43) $$\frac{1}{S_{\mathbf{k}}-2\pi ik_l}=\left\{-\frac{1}{2\pi ik_l}+t\,e^{2\pi ik_lt}\right\}*,\ l=1, 2, \ldots, m,$$

where by $\{\ \}*$ it is denoted that the corresponding function is considered as a Bernoulli convolution operator with this function as kernel.

Proof. Assume that $S_{\mathbf{k}}-2\pi ik_l$ is a divisor of zero in $\mathsf{M}_{\mathbf{k}}$. Then, there should exist a function $\varphi\in\mathscr{C}_{\mathbf{k}}$, $\varphi\neq0$, such that $(S_{\mathbf{k}}-2\pi ik_l)\varphi=0$. Multiplying by $L_{\mathbf{k}}$ we get $\varphi-2\pi ik_lL_{\mathbf{k}}\varphi=0$, i. e. that φ is an eigenfunction of $L_{\mathbf{k}}$ with eigenvalue $1/(2\pi ik_l)$. Hence, $\varphi(t)=Ae^{2\pi ik_lt}$ with $A\neq0$. This is a contradiction, since $e^{2\pi ik_lt}$ is not a function of $\mathscr{C}_{\mathbf{k}}$.

It remains (42) to be verified. By Lemma 12, it is equivalent to

$$\frac{1}{S_{\mathbf{k}}-2\pi ik_l}=\left\{-\frac{1}{2\pi ik_l}+t\,e^{2\pi ik_lt}+\frac{1}{2}\,e^{2\pi ik_lt}+\sum_{\substack{j=1\\ j\neq l}}^{m}\frac{e^{2\pi ik_jt}}{2\pi i(k_l-k_j)}\right\}*.$$

Denoting the right-hand side by φ_l, this equation is equivalent to

$$\varphi_l-2\pi ik_lL_{\mathbf{k}}\varphi_l=L_{\mathbf{k}}$$

in $\mathsf{M}_{\mathbf{k}}$. Therefore, taking into account (39) and (40), we should verify the identity

$$\varphi_l-2\pi ik_lL_{\mathbf{k}}\varphi_l=t-\frac{1}{2}-\sum_{j=1}^{m}\frac{1}{2\pi ik_j}\,e^{2\pi ik_jt}$$

in $\mathscr{C}_{\mathbf{k}}$. This can be settled by a direct check.

The partial fractions $1/(S_{\mathbf{k}}-\lambda)^n$ in the non-resonance case can be represented as

(44) $$\frac{1}{(S_{\mathbf{k}}-\lambda)^n}=\left\{\frac{(-1)^n}{\lambda^n}+\frac{1}{(n-1)!}\,\frac{\partial^{n-1}}{\partial\lambda^{n-1}}\left(\frac{e^{\lambda t}}{e^{\lambda}-1}\right)\right\}*,$$

as it had already been proved in 2.5.1. □

Theorem 2.5.5 (Grozdev [63]). *Let n be an arbitrary positive integer. Then the representation*

(45) $$\frac{1}{(S_{\mathbf{k}}-2\pi ik_l)^n}=\left\{\frac{(-1)^n}{(2\pi ik_l)^n}+\frac{e^{2\pi ik_lt}}{n!}B_n(t)\right\}*$$

holds. Here $B_n(t)$ is the n-th Bernoulli polynomial, and k_l is an integer, involved in the multi-subscript $\mathbf{k}$.

Proof. Theorem 2.5.4 settles only the case $n=1$. The general proposition follows by an inductive argument. Let us suppose that (45) is true for some $n\geq1$. Then for $\lambda=2\pi ik_l$

$$\frac{1}{(S_{\mathbf{k}}-\lambda)^{n+1}}=\frac{1}{S_{\mathbf{k}}-\lambda}\cdot\frac{1}{(S_{\mathbf{k}}-\lambda)^{n}}$$

$$=\left[\left(-\frac{1}{\lambda}+e^{\lambda t}B_1(t)\right)*\left(\frac{(-1)^n}{\lambda^n}+\frac{e^{\lambda t}}{n!}B_n(t)\right)\right]*.$$

Since the constant function $\{1\}$ is the unit of the convolution $*$, then for the convolution product in the brackets we get

$$\left(-\frac{1}{\lambda}+e^{\lambda t}B_1(t)\right)*\left(\frac{(-1)^n}{\lambda^n}+\frac{e^{\lambda t}}{n!}B_n(t)\right)=\frac{(-1)^{n+1}}{\lambda^{n+1}}$$

$$-\frac{1}{\lambda n!}e^{\lambda t}B_n(t)+\frac{(-1)^n}{\lambda^n}e^{\lambda t}B_1(t)+\frac{1}{n!}(e^{\lambda t}B_1(t))*(e^{\lambda t}B_n(t)).$$

But

$$e^{\lambda t}B_1(t)*e^{\lambda t}B_n(t)=e^{\lambda t}B_1(t)\int_0^1 e^{\lambda\tau}B_n(\tau)d\tau+e^{\lambda t}B_n(t)\int_0^1 e^{\lambda t}B_1(\tau)d\tau$$

$$-e^{\lambda t}\left[\int_0^t B_1(t-\tau)B_n(\tau)d\tau+\int_t^1 B_1(1+t-\tau)B_n(\tau)d\tau\right].$$

By an easy check it is seen that

$$\int_0^1 e^{\lambda\tau}B_m(\tau)d\tau=\frac{(-1)^{m-1}m!}{\lambda^m},\qquad m=1,2,\ldots.$$

Using the characteristic properties $B'_{m+1}(t)=(m+1)B_m(t)$, $B_0(t)\equiv 1$ and $\int_0^1 B_m(\tau)d\tau=0$ for $m=1,2,\ldots$, and integrating by parts, we get easily

$$\int_0^t B_1(t-\tau)B_n(\tau)d\tau+\int_t^1 B_1(1+t-\tau)B_n(\tau)d\tau=-\frac{B_{n+1}(t)}{n+1},$$

thus proving (45) by induction. □

Corollary. If $k\neq 0$ is an integer, and the multi-subscript $\mathbf{k}$ consists of k only, i. e. $\mathbf{k}=(k)$, then

$$\frac{1}{(S_{\mathbf{k}}-2\pi ik)^n}=\frac{(-1)^n}{(2\pi ik)^n}+\frac{e^{2\pi ikt}}{n!}B_n(t). \tag{46}$$

In order to establish (46), using (45), one should show only that the right-hand side of (46) is a function of $\mathscr{C}_k$.

Now we are ready to propose a convolutional algorithm for the solution of the boundary value problem

(47) $$P\left(\frac{d}{dt}\right)x=f,\quad x_0=\gamma_0,\quad x_0^{(k)}=x^{(k-1)}(1)-x^{(k-1)}(0)=\gamma_k,$$

$$k=1, 2, \ldots, n-1,$$

in the resonance case, when the polynomial $P(\lambda)=a_0\lambda^n+a_1\lambda^{n-1}+\ldots+a_{n-1}\lambda+a_n$ with $a\neq 0$ is such that $P(2\pi ik_l)=0$ for the integers k_l, $l=1, 2, \ldots, m$, which are different from 0. Let us remind that we have used the notation $x_0=\int_0^1 x(\tau)d\tau$. As in the non-resonance case, considered in 2.3.1, problem (47) can be reduced to the single equation

(48) $$P(S)x=f+Q(S)$$

in the multiplier quotients ring M of the Bernoulli convolution algebra $\mathscr{C}([0, 1])$, where

$$Q(S)=\sum_{j=1}^{n}\left(\sum_{k=j}^{n}a_{n-k}\gamma_{k-j}\right)S^j.$$

But in the resonance case we are dealing with, equation (48) has not always solution in M, since $P(S)$ is a divisor of zero in M. If it has such a solution, it cannot be unique, since if $a_1, a_2, \ldots, a_m$ are arbitrarily chosen in C, then the element $y=a_1\{e^{2\pi ik_1t}\}+\ldots+a_n\{e^{2\pi ik_mt}\}$ is a solution of the equation $P(S)y=0$.

In order to obtain the necessary conditions for the solvability of (48) in M, let us multiply (48) by $\{e^{2\pi ik_lt}\}$, $l=1, 2, \ldots, m$. Since $P(S)\{e^{2\pi ik_lt}\}=P(2\pi ik_l)e^{2\pi ik_lt}=0$ and $Q(S)\{e^{2\pi ik_lt}\}=Q(2\pi ik_l)\{e^{2\pi ik_lt}\}$, then we get

(49) $$\chi_{k_l}(f)+Q(2\pi ik_l)=0,\qquad l=1, 2, \ldots, m$$

as necessary conditions for solvability of (48).

Lemma 14. *If conditions* (49) *are fulfilled, then boundary value problem* (47) *has a unique solution in* $\mathscr{C}_{\mathrm{k}}$.

Proof. By means of purely algebraic considerations, using (48), it is easy to see that the boundary value problem (47) is equivalent to the integral equation

(50) $$P^*(L)x=L^nf+Q^*(L)$$

with $P^*(\lambda)=\lambda^nP(1/\lambda)$ and $Q^*(\lambda)=\lambda^nQ(1/\lambda)$. It is easy to see that (49) are equivalent to the condition the right-hand side of (50) to be a function of $\mathscr{C}_{\mathrm{k}}$, i. e. $L^nf+Q^*(L)\in\mathscr{C}_{\mathrm{k}}$. Applying the projector $P_{\mathrm{k}}=e_k*$ to (50), we obtain

$$P^*(L_{\mathrm{k}})\tilde{x}=L_{\mathrm{k}}^n\tilde{f}+Q^*(L_{\mathrm{k}})$$

with $\tilde{x}=P_{\mathrm{k}}x$ and $\tilde{f}=P_{\mathrm{k}}f$. This, considered as an equation in $\mathscr{C}_{\mathrm{k}}$, can be written in the form $P(S_{\mathrm{k}})\tilde{x}=\tilde{f}+Q(S_{\mathrm{k}})$. By Theorem 2.5.4, $P(S_{\mathrm{k}})$ is a non-divisor of zero in M_{k} and hence

(51) $$\tilde{x}=\frac{1}{P(S_{\mathrm{k}})}\tilde{f}+\frac{Q(S_{\mathrm{k}})}{P(S_{\mathrm{k}})}$$

is a solution of the above equation in $\mathsf{M}_{\mathbf{k}}$. It remains to be seen that (51) gives a true solution of boundary value problem (47) in $\mathscr{C}_{\mathbf{k}}$. Indeed, $1/P(S_{\mathbf{k}})$ can be decomposed into partial fractions, and each of them is a convolution operator in $\mathscr{C}_{\mathbf{k}}$. Hence $[1/P(S_{\mathbf{k}})]\tilde{f}$ is a function of $\mathscr{C}_{\mathbf{k}}$. As for the second term $Q(S_{\mathbf{k}})/P(S_{\mathbf{k}})$ it is sum of a constant, i. e. numerical multiple of the unit $e_{\mathbf{k}}$ of $\mathscr{C}_{\mathbf{k}}$ and sum of partial fractions, each of which, by Theorems 2.3.1 and 2.3.5, lies in $\mathscr{C}_{\mathbf{k}}$.

One can verify in the same way as in the non-resonance case that $\tilde{x} \in \mathscr{C}_{\mathbf{k}}$, given by (51), is a solution of boundary value problem (47). □

In order to get all the solutions of the boundary value problem (47), it remains only to find all the solutions of the homogeneous boundary value problem

(52) $$P\left(\frac{d}{dt}\right)y=0; \quad y_0^{(j)}=0, \quad j=0, 1, \dots, n-1.$$

This problem is equivalent to the equation $P(S)y=0$. All the solutions of this equation are contained in the formula

(53) $$y=\{c_1 e^{2\pi i k_1 t}+ \dots + c_m e^{2\pi i k_m t}\},$$

where $c_1, c_2, \dots, c_m$ are arbitrary constants.

Now let us summarize the algorithm for solving (47).

First of all, conditions (49) should be verified. If they are fulfilled, then we proceed in the following way. Denote by $\lambda_l = 2\pi i k_l$, $l=1, 2, \dots, m$, the "resonance zeros" of $P(\lambda)$, and by $\lambda_{m+1}, \dots, \lambda_r$ the non-resonance ones. By $\varkappa_j$, $j=1, 2, \dots, r$, we denote the multiplicity of the zero λ_j. Next we find the partial fractions decompositions

$$\frac{1}{P(S_{\mathbf{k}})} = \sum_{l=1}^{r} \sum_{j=1}^{\varkappa_l} \frac{A_{lj}}{(S_{\mathbf{k}}-\lambda_l)^j}$$

and

$$\frac{Q(S_{\mathbf{k}})}{P(S_{\mathbf{k}})} = \gamma_0 + \sum_{l=1}^{r} \sum_{j=1}^{\varkappa_l} \frac{A_{lj}^0}{(S_{\mathbf{k}}-\lambda_l)^j}$$

with appropriate constants A_{lj}, A_{lj}^0 and γ_0.

Finally, we write down the solution of (47) in the form

(54) $$\tilde{x} = g * f + g_0 + \sum_{l=1}^{m} a_l e^{2\pi i k_l t} + \gamma_0$$

with

$$g(t) = \sum_{l=1}^{m} \sum_{j=1}^{\varkappa_l} A_{lj} \left\{ \frac{(-1)^j}{(2\pi i k_l)^j} + \frac{e^{2\pi i k_l t}}{j!} B_j(t) \right\}$$

$$+ \sum_{l=m+1}^{r} \sum_{j=1}^{\varkappa_l} A_{lj} \left\{ \frac{(-1)^j}{\lambda_l^j} + \frac{1}{(j-1)!} \frac{\partial^{j-1}}{\partial \lambda_l^{j-1}} \left(\frac{e^{\lambda_l t}}{e^{\lambda_l}-1} \right) \right\}$$

and with g_0 of the same form as g, but with A^0_{lj} instead of A_{lj}, and with arbitrary constants $a_1, \ldots, a_m$.

We can state this as

Theorem 2.5.6. *Let $\lambda_l = 2\pi i k_l$ with integers k_l, $l=1, 2, \ldots, m$, be different zeros of an n-th degree polynomial $P(\lambda) = \alpha_0\lambda^n + \alpha_1\lambda^{n-1} + \ldots + \alpha_{n-1}\lambda + \alpha_n$ with multiplicities $\varkappa_l$, and let $P(2\pi i k) \neq 0$ for all integers $k \neq k_l$, $k \neq 0$. Let the other zeros of $P(\lambda)$ be $\lambda_{m+1}, \ldots, \lambda_r$, with the multiplicities $\varkappa_l$, correspondingly. Then the boundary value problem*

$$P\left(\frac{d}{dt}\right)x = f, \quad \int_0^1 x(\tau)d\tau = \gamma_0, \quad x^{(k-1)}(1) - x^{(k-1)}(0) = \gamma_k,$$

$k=1, 2, \ldots, n-1$, has a solution in $\mathscr{C}([0, 1])$ iff the conditions

$$\int_0^1 (1 - e^{-2\pi i k_l t}) f(t)dt + \sum_{j=1}^{n} \left(\sum_{k=j}^{n} \alpha_{n-k}\gamma_{k-j} \right) (2\pi i k_l)^j = 0, \quad l = 1, 2, \ldots, m,$$

are fulfilled. In this case the problem has ∞^m solutions, and they are all given by (54).

Instead of the segment [0, 1] one can consider $(-\infty, \infty)$ and seek 2π-periodic solutions of an ordinary linear differential equation $P(d/dt)y = f$ or systems of such equations, provided f is 2π-periodic. It is easy to conceive that by means of some obvious modifications the above algorithms can be extended to such problems too. In Kaplan [68] this problem is treated by means of the *finite Fourier transform,* but only in the non-resonance case. It seems that the direct algebraic approach, based on the evident modification of the Bernoulli convolution (5), has some advantages as compared with the use of the finite Fourier transform.

CHAPTER 3

CONVOLUTIONS CONNECTED WITH SECOND-ORDER LINEAR DIFFERENTIAL OPERATORS

In contrast to the differentiation operator playing a basic role in analysis, the linear differential operators of the second order are of most essential importance in mathematical physics. Here the classical Sturm-Liouville boundary value problem should be mentioned. But in some modern problems the local boundary conditions, as those in the Sturm-Liouville problem, are inadequate and the need of a general treatment of non-local boundary value conditions arises.

In this chapter an attempt at extending the convolutional approach to boundary value problems connected with second-order linear differential operators is made. We are trying to embrace both local and non-local boundary value conditions. A number of new convolutions is proposed. By specialization, convolutions for some commonly used finite integral transformations are found in explicit form, e. g. for Sturm-Liouville and for Hankel finite integral transformations. By forming tensor products of such convolutions, new Duhamel-type representations of some problems of mathematical physics can be found. It is worth mentioning that such Duhamel representations can be used not only for theoretical purposes, but as an alternative of the finite difference methods in numerical calculation of the solutions of many well-known problems of mathematical physics.

3.1. CONVOLUTIONS OF RIGHT INVERSE OPERATORS OF THE SQUARE OF THE DIFFERENTIATION

The square of the differentiation plays much the same role in the theory of linear differential operators of the second order, as the differentiation operator in that of the linear differential operators of the first order. But here the sutuation is far more complicated. One reason for such complexity can be seen in the fact that whereas an arbitrary right inverse operator of the differentiation operator can be expressed by a single linear functional, for expressing an arbitrary right inverse operator of the square of the differentiation, two such linear functionals are needed. We still cannot propose convolutions for such general right inverses in explicit form. But most of the known boundary value problems of mathematical physics can be embraced into a scheme, using only one arbitrary linear functional.

The right inverses of the square of the differentiation for which we can propose convolutions in explicit form are determined by one Sturm-Liouville boundary value condition and one quite general boundary value condition, depending on an arbitrary linear functional.

3.1.1. A convolution connected with the square of the differentiation and depending on an arbitrary linear functional. Here we consider a segment $[0, T]$ and the space $\mathscr{C}([0, T])$ of the complex-valued continuous functions on $[0, T]$. For a unification, let us take $T=1$, and later on to consider only the space $\mathscr{C}=\mathscr{C}([0, 1])$.

Theorem 3.1.1. *Let $\Phi\colon \mathscr{C}\to \mathbf{C}$ be a non-zero linear functional on $\mathscr{C}$. Then the operation*

$$(f*g)(t)=\Phi_x\{k(x,t)\}, \tag{1}$$

with

$$k(x,t)=\frac{1}{2}\int_t^x f(x+t-\tau)g(\tau)d\tau-\frac{1}{2}\int_{-t}^x f(|x-t-\tau|)g(|\tau|)\operatorname{sgn}((x-t-\tau)\tau)d\tau$$

is a bilinear, commutative and associative operation in $\mathscr{C}$. It is separately continuous and annihilators-free too.

Proof. If we introduce the odd continuations $f_0(t)=f(|t|)\operatorname{sgn} t$ and $g_0(t)=g(|t|)\operatorname{sgn} t$ of $f(t)$ and $g(t)$, the expression $k(x, t)$ under the functional sign can be written in a simpler form:

$$k(x,t)=\frac{1}{2}\int_t^x f(x+t-\tau)g(\tau)d\tau-\frac{1}{2}\int_{-t}^x f_0(x-t-\tau)g_0(\tau)d\tau. \tag{2}$$

The bilinearity and commutativity of (1) are evident. Before proving the associativity of (1), let us prove that it is separately continuous. But this is evident from the inequality

$$\|f*g\|_C\leqq\|\Phi\|.\|f\|_C\cdot\|g\|_C, \tag{3}$$

where $\|\ \|_C$ denotes the supremum-norm in $\mathscr{C}$, and $\|\Phi\|$ — the norm of Φ in $\mathscr{C}$. But for the proof of the associativity of (1) we shall need an estimate of the form

$$\|f*g\|_1\leqq A\,\|f\|_1.\|g\|_1 \quad \text{with} \quad A=\text{const} \tag{4}$$

in $\mathscr{C}([0, 1])$, where $\|f\|_1$ denotes the integral norm $\|f\|_1=\int_0^1|f(t)|\,dt$. Here we drop the simple and routine calculation leading to (4).

Now let us prove the associativity of (1). First we verify the associativity relation for functions of the form $f_\lambda(t)=\sin\lambda t/\lambda$ with $\lambda\neq 0$. We have

$$(f_\lambda*f_\mu)(t)=\frac{E(\mu)f_\lambda(t)-E(\lambda)f_\mu(t)}{\lambda^2-\mu^2}, \tag{5}$$

if $\lambda^2\neq\mu^2$. Here we have denoted $E(z)=\Phi_x\{\cos zx\}$. Then, using (5), we get easily

(6) $$(f_\lambda * f_\mu) * f_\nu = \frac{E(\mu)E(\nu)}{(\lambda^2-\mu^2)(\lambda^2-\nu^2)} f_\lambda + \frac{E(\nu)E(\lambda)}{(\mu^2-\lambda^2)(\mu^2-\nu^2)} f_\mu + \frac{E(\lambda)E(\mu)}{(\nu^2-\lambda^2)(\nu^2-\mu^2)} f_\nu.$$

Due to the symmetry of this expression with respect to λ, μ, ν, the associativity relation

(7) $$(\{\sin \lambda t\} * \{\sin \mu t\}) * \{\sin \nu t\} = \{\sin \lambda t\} * (\{\sin \mu t\} * \{\sin \nu t\})$$

is established. Now we shall differentiate (7) with respect to the parameters λ, μ, ν. This is allowed, since Φ is a continuous linear functional. Differentiating $2l+1$, $2m+1$ and $2n+1$ times on λ, μ and ν correspondingly, we get

$$(\{(-1)^l t^{2l+1} \cos \lambda t\} * \{(-1)^m t^{2m+1} \cos \mu t\}) * \{(-1)^n t^{2n+1} \cos \nu t\}$$
$$= \{(-1)^l t^{2l+1} \cos \lambda t\} * (\{(-1)^m t^{2m+1} \cos \mu t\} * \{(-1)^n t^{2n+1} \cos \nu t\}).$$

For $\lambda, \mu, \nu \to 0$ we get

(8) $$(\{t^{2l+1}\} * \{t^{2m+1}\}) * \{t^{2n+1}\} = \{t^{2l+1}\} * (\{t^{2m+1}\} * \{t^{2n+1}\})$$

for $l, m, n = 0, 1, 2, \dots$. Hence the associativity relation $(f*g)*h = f*(g*h)$ holds for odd polynomials

$$f(t) = \sum_{l=0}^{p} a_{pl} t^{2l+1}, \quad g(t) = \sum_{m=0}^{q} b_{qm} t^{2m+1}, \quad h(t) = \sum_{n=0}^{r} c_{rn} t^{2n+1}.$$

By means of the Weierstrass approximation theorem it can easily be seen that each function $\varphi(t) \in \mathscr{C}([0, 1])$, with $\varphi(0) = 0$, can be uniformly approximated by odd polynomials. Thus, using the separate continuity of (1) with respect to the uniform convergence, we may take it for granted that the associativity relation is proved for functions $f, g, h \in \mathscr{C}([0, 1])$ with $f(0) = g(0) = h(0) = 0$. Then in order to prove the associativity in $\mathscr{C}([0, 1])$, we should verify the identities

(9) $$(\{1\} * g) * h = \{1\} * (g * h) = (\{1\} * h) * g$$

and

(10) $$(\{1\} * \{1\}) * h = \{1\} * (\{1\} * h),$$

assuming that g and h are functions from $\mathscr{C}([0, 1])$, such that $g(0) = h(0) = 0$. Identity (7) is convenient to be verified at first for functions of the form $g(t) = \sin \lambda t$ and $h(t) = \sin \mu t$. Denoting $F(z) = \Phi_x\{\sin zx\}$ and $E(\lambda) = \Phi_x\{\cos \lambda x\}$, we get

(11) $$\{1\} * \{\sin \lambda t\} = \frac{1}{\lambda}[E(\lambda) \cos \lambda t + F(\lambda) \sin \lambda t - E(\lambda)]$$

and

(12) $$\{\cos \lambda t\} * \{\sin \mu t\} = \frac{\lambda}{\lambda^2-\mu^2} F(\lambda) \sin \mu t - \frac{\mu}{\lambda^2-\mu^2}[E(\mu) \cos \mu t + F(\mu) \sin \mu t - E(\mu) \cos \lambda t].$$

Then, with a direct check, using only formulas (11) and (12), we see that

$$(13)\qquad (\{1\}*\{\sin \lambda t\})*\{\sin \mu t\}=\{1\}*(\{\sin \lambda t\}*\{\sin \mu t\}) =(\{1\}*\{\sin \mu t\})*\{\sin \nu t\},$$

at least for $\lambda^2 \neq \mu^2$. Further, by differentiating (13) on λ ane μ and letting $\lambda, \mu \to 0$, we get

$$(14)\quad (\{1\}*\{t^{2l+1}\})*\{t^{2m+1}\}=\{1\}*(\{t^{2l+1}\}*\{t^{2m+1}\})=(\{1\}*\{t^{2m+1}\})*\{t^{2l+1}\}$$

for $l, m=0, 1, 2, \ldots$. Now, using again uniform polynomial approximation, we prove (9).

In proving (4) it is easier to use the estimate (4). We shall use the fact that the set of the odd polynomials is dense in $\mathscr{L}([0, 1])$ with respect to the integral norm $\|f\|_1=\int_0^1 |f(t)|\,dt$. If f, g and h are arbitrary functions from $\mathscr{C}([0, 1])$, then let $f_n(t)$, $g_n(t)$, $h_n(t)$ be odd polynomial sequences converging to $f(t)$, $g(t)$, $h(t)$ correspondingly with respect to the integral norm, i. e. such that $\|f_n-f\|_1\to 0$, $\|g_n-g\|_1\to 0$, $\|h_n-h\|_1\to 0$ for $n\to\infty$. The estimate (4) shows that from $(f_n*g_n)*h_n=f_n*(g_n*h_n)$ it follows $(f*g)*h=f*(g*h)$, thus proving the associativity in the general case of $\mathscr{C}([0, 1])$.

It remains to prove that (1) is an annihilators-free operation in $\mathscr{C}([0, 1])$. To this end it is sufficient to show that there exists a non-trivial non-divisor of zero in $\mathscr{C}([0, 1])$. Since it is assumed that Φ is a non-zero functional, then the function $E(\lambda)=\Phi_x\{\cos \lambda x\}$ cannot be identically zero. Indeed, if it were $\Phi_x\{\cos \lambda x\}=0$ for all $\lambda \in \mathbb{C}$, then by an even number of differentiations on λ, we could get $\Phi_x\{(-1)^n x^{2n}\cos \lambda x\}=0$, and letting $\lambda\to 0$, we would get $\Phi_x\{x^{2n}\}=0$ for $n=0, 1, \ldots$. Hence $\Phi(f(t))=0$ for each even polynomial of t. Hence, by approximation, we would get $\Phi(f)=0$ in $\mathscr{C}([0, 1])$. □

Now, let $\lambda\in\mathbb{C}$ be such that $E(\lambda)\neq 0$. Let us consider the convolutional operator

$$(15)\qquad L_\lambda f(t)=\left\{\frac{\sin \lambda t}{\lambda E(\lambda)}\right\}*f(t).$$

Developed by (1) it can be given the form

$$(16)\quad L_\lambda f(t)=\frac{1}{\lambda}\int_0^t \sin \lambda(t-\tau)\,f(\tau)d\tau-\frac{\sin \lambda t}{\lambda E(\lambda)}\Phi_x\left\{\int_0^x \cos \lambda(x-\tau)\,f(\tau)d\tau\right\}.$$

Written in this form, it is not difficult to see that L_λ is a right inverse operator of the differential operator $D_\lambda=D^2+\lambda^2$, such that $y=L_\lambda f$ is the solution of the elementary boundary value problem

$$(17)\qquad y''+\lambda^2 y=f(t),\quad y(0)=0,\quad \Phi(y')=0.$$

The condition $E(\lambda)\neq 0$ ensures that (17) has a unique solution. Therefore the function $\sin \lambda t/\lambda E(\lambda)$ is not a divisor of zero of the operation (1), since if $\{\sin \lambda t/\lambda E(\lambda)\} * f=0$, then $L_\lambda f=0$. Applying the operator D_λ, we get $D_\lambda L_\lambda f=0=f$.

Now we begin to study the functional properties of (1). To exhibit them, let us transform the expression under the functional sign.

Lemma 1. *For $f, g \in \mathscr{C}$ the identity*

$$k(x, t)=-\int_0^{|x-t|} f(|x-t|-\tau)g(\tau)d\tau+\frac{1}{2}\int_t^x f(x+t-\tau)g(\tau)d\tau + \frac{1}{2}\int_{-t}^x f(|x-t-\tau|)g(|\tau|)d\tau, \tag{18}$$

where $k(x, t)$ is given by (2), *holds.*

Proof. It is easy to see that (18) is equivalent to the equation

$$\int_{-t}^x f(|x-t-\tau|)g(|\tau|)d\tau+\int_{-t}^x f(|x-t-\tau|)g(|\tau|)(\operatorname{sgn}(x-t-\tau)\tau)d\tau = 2\int_0^{|x-t|} f(|x-t|-\tau)g(\tau)d\tau. \tag{19}$$

Then we shall prove (19). Let us first consider the case $x \geqq t$. Then

$$\int_{-t}^x f(|x-t-\tau|)g(|\tau|)d\tau=\int_{-t}^0+\int_0^{x-t}+\int_{x-t}^x=\int_0^t f(x-t+\tau)g(\tau)d\tau + \int_0^{x-t} f(x-t-\tau)g(\tau)d\tau+\int_{x-t}^x f(\tau+t-x)g(\tau)d\tau$$

and

$$\int_{-t}^x f(|x-t-\tau|)g(|\tau|)\operatorname{sgn}((x-t-\tau)\tau)d\tau=-\int_0^t f(x-t-\tau)g(\tau)d\tau + \int_0^{x-t} f(x-t-\tau)g(\tau)d\tau-\int_{x-t}^x f(\tau+t-x)g(\tau)d\tau,$$

and (19) holds. The case $x \leqq t$ is treated in the same way. □

By means of Lemma 1 and using the functional properties of the convolution

$$(f \,\widehat{*}\, g)(t)=\Phi_x\left\{\int f(x+t-\tau)g(\tau)d\tau\right\}, \tag{20}$$

proved in sect. 2.1, analogous properties of (1) can be established.

Theorem 3.1.2. *If $f, g \in \mathscr{C}([0, 1])$, and if f is a function with bounded variation on $[0, 1]$, and if $f(0)=0$, then $f * g \in \mathscr{C}^1([0, 1])$.*

Proof. According to the results of sect. 2.1, the functions

$$h_2(t)=\Phi_x\left\{\int_t^x f(x+t-\tau)g(\tau)d\tau\right\},\ h_3(t)=\Phi_x\left\{\int_{-t}^x f(|x-t-\tau|)g(|\tau|)d\tau\right\}$$

are in $\mathscr{C}^1([0, 1])$. We shall show that under the assumption $f(0)=0$, the function

$$h_1(t)=\Phi_x\left\{\int_0^{|t-x|} f(|t-x|-\tau)g(\tau)d\tau\right\}$$

belongs to the space $\mathscr{C}^1([0, 1])$ too. Indeed, as we had shown in sect. 1.1, the Duhamel convolution of a function $f(t)$ with a bounded variation and a continuous function $g(t)$ is a smooth function, and when $f(0)=0$, the relation

$$\frac{d}{dt}\int_0^t f(t-\tau)g(\tau)d\tau=\int_0^t g(t-\tau)df(\tau)$$

holds. If $f(0)=0$, then the function $h_1(t)$ is differentiable too, and

$$h_1'(t)=\Phi_x\left\{\operatorname{sgn}(t-x)\int_0^{|t-x|} g(|t-x|-\tau)df(\tau)\right\}.$$

The expression in the right-hand side is well defined, since the function under the functional sign is a continuous function of the two variables t and x. It is clear that $h_1(t)\in\mathscr{C}^1([0, 1])$. Taking into account the expression for the derivatives of (20), obtained in sect. 2.1, in our case we get

$$(21)\qquad (f*g)'(t)=\Phi_x\left\{\operatorname{sgn}(t-x)\int_0^{|x-t|} g(|x-t|-\tau)df(\tau)\right.$$

$$\left.-\int_t^x g(x+t-\tau)df(\tau)+\int_{-t}^x g(|x-t-\tau|)df(|\tau|)\right\}.\ \square$$

A $\mathscr{L}^1$-version of this theorem can be stated:

If $f, g \in \mathscr{L}^1([0, 1])$, and if f is equal almost everywhere in $[0, 1]$ to a function with bounded variation, then the function $f*g$, defined by (1), is absolutely continuous on $[0, 1]$.

We do not give a proof of this theorem, since we do not use it later.

3.1.2. Convolutions of the first kind right inverses of d^2/dt^2. As we have already said, for unification of the following considerations, we take as basic the space $\mathscr{C}=\mathscr{C}([0, 1])$. Here we consider the class of the right inverse operators of the square of the differentiation in $\mathscr{C}$, defined by a boundary value condition of the first kind in the point $t=0$ and by an arbitrary boundary value condition, determined by a linear functional N in the space $\mathscr{C}^1([0, 1])$. If $L: \mathscr{C}\to\mathscr{C}^2$ is such a right inverse operator of (d^2/dt^2), then the function $y=Lf$ should be the solution of the elementary boundary value problem

$$y''=f, \quad y(0)=0, \quad N(y)=0. \tag{22}$$

It is assumed that N is a continuous linear functional on $\mathscr{C}^1([0, 1])$, such that the problem (22) has a solution for all $f\in\mathscr{C}([0, 1])$. The continuity of N should be understood with respect to the $\mathscr{C}^1$-norm

$$\|f\|_{C^1}=\max_{t\in[0,1]}\{|f(t)|+|f'(t)|\}. \tag{23}$$

It is easy to see that for the solvability of (22) it is necessary to assume $N\{x\}\neq 0$. There is no loss of generality to assume that

$$N\{x\}=1. \tag{24}$$

If, as earlier, we use l for the Volterra integration operator

$$lf(t)=\int_0^t f(\tau)d\tau, \tag{25}$$

then the right inverse operator L of d^2/dt^2, defined by the boundary value problem (22), has a representation of the form

$$Lf(t)=l^2f(t)-N(l^2f)t. \tag{26}$$

The defining projector of L has the form

$$Ff(t)=f-Lf''=f(0)[1-N\{1\}t]+N(f)t \tag{27}$$

and it can be considered as defined on the whole space $\mathscr{C}^1([0, 1])$.

Our main aim here is to find a convolution of L in an explicit form. This problem is solved in Dimovski [37].

Theorem 3.1.3. *The operation*

$$(f*g)(t)=-(N\circ l)_x\left\{\frac{1}{2}\int_t^x f(x+t-\tau)g(\tau)d\tau\right. \tag{28}$$

$$\left.-\frac{1}{2}\int_{-t}^x f(|x-t-\tau|)g(|\tau|)\operatorname{sgn}((x-t-\tau)\tau)\,d\tau\right\}$$

is a convolution of the operator (26) *in* $\mathscr{C}([0, 1])$, *and has the representation*

$$Lf=\{t\}*f \tag{29}$$

in it.

Proof. Since N is a linear functional in $\mathscr{C}^1$, then $\Phi=-N\circ l$ is a linear functional in $\mathscr{C}$. Then operation (28) has the form (1) and by Theorem 3.1.1 it is a bilinear, commutative and associative operation in $\mathscr{C}$. In order to assert that (28) is a convolution of L in $\mathscr{C}$, we should prove representation (29). This can be done by a direct check.

Even for smooth functions f, g operation (28) does not always give a differentiable function $f*g$. In order to ensure $f*g$ to be differentiable, one should assume $f(0)=0$. Let us introduce the invariant subspace

(30) $$\mathscr{C}_0=\mathscr{C}_0([0, 1])=\{f(t)\in\mathscr{C}([0, 1]), f(0)=0\}$$

of the operator L. If $f, g\in\mathscr{C}$, then it is easy to see that $(f*g)(0)=0$, i. e. $*:\mathscr{C}\times\mathscr{C}\to\mathscr{C}_0$. Hence $*$ is a convolution of L in $\mathscr{C}_0$ too. □

Theorem 3.1.4. *The space $\mathscr{C}_0^2$ of twice continuously differentiable functions $f(t)$ in $\mathscr{C}([0, 1])$ with $f(0)=0$ is an ideal in the convolutional algebra $\mathscr{C}$ with multiplication* (28), *and for $f\in\mathscr{C}_0^2$ and $g\in\mathscr{C}$ the differentiation formula*

(31) $$(f*g)''=f''*g+N(f)g$$

holds.

Proof. If $f\in\mathscr{C}_0^2$, then by formula (27) we get

$$f=Lf''+N(f)t.$$

Therefore, $f*g=(Lf'')*g+N(f)\{t\}*g=L\{f''*g+N(f)g\}$ and hence $f*g\in\mathscr{C}_0^2$ and $(f*g)''=f''*g+N(f)g$. □

Example 1. If $N(f)=f'(0)$, then we get $Lf=l^2f$ and (28) gives

$$(f*g)(t)=\int_0^t f(t-\tau)g(\tau)d\tau,$$

i. e. the usual Duhamel convolution, which could be expected.

Example 2. Let $N(f)=f(1)$. Then $y=Lf$ is the solution of the two-point boundary value problem

(32) $$y''=f,\ y(0)=y(1)=0.$$

In this case

(33) $$Lf=l^2f(t)-t\int_0^1(1-\tau)f(\tau)d\tau.$$

The defining projector F of L is the operator

(34) $$Ff(t)=(I-LD)f(t)=(1-t)f(0)+tf(1).$$

By specialization of operation (28), as a convolution of L in $\mathscr{C}([0, 1])$ we get

(35) $$f*g(t)=-\int_0^1\left[\frac{1}{2}\int_t^x f(x+t-\tau)g(\tau)d\tau\right.$$

$$-\frac{1}{2}\int_{-t}^{x} f(|x-t-\tau|)g(|\tau|)\,\mathrm{sgn}\,((x-t-\tau)\tau)d\tau\Bigg]dx.$$

The operator (33) presents some interest from the point of view of the analysis, since it is closely connected with Lidstone's polynomials (see Whittaker [106]).

Definition. The polynomial system $\{A_n(t)\}_{n=1}^{\infty}$, determined recurrently by $A_n''(t)=A_{n-1}(t)$, $A_0(t)=t$ and by the boundary value conditions $A_n(0)=A_n(1)=0$ for $n\geq 1$ is said to be the *Lidstone polynomial system.*

Due to asymmetry, caused by the choice $A_0(t)=t$, these polynomials can be named more exactly left Lidstone polynomials. The right Lidstone polynomials can be defined recurrently by $\tilde{A}_n''(t)=\tilde{A}_{n-1}(t)$, the initial choice $\tilde{A}_0(t)=1-t$ and by the same boundary value conditions $\tilde{A}_n(0)=\tilde{A}_n(1)=0$, $n\geq 1$.

Lemma 2. *For each integer $n\geq 0$, identically*

$$\tilde{A}_n(t)=A_n(1-t). \tag{36}$$

Proof. Let us denote $A_n(t)=A_n(1-t)$, $n\geq 0$. Obviously, $A_n''(t)=A_n''(1-t)=A_{n-1}(1-t)=A_{n-1}(t)$ and $A_0(t)=1-t$. The boundary value conditions $A_n(0)=A_n(1)=0$ are fulfilled. Hence $A_n(t)=\tilde{A}_n(t)$.

The generalized Taylor formula (see 1.3.4) for the right inverse L of d^2/dt^2 takes the form

$$f(t)=\sum_{k=0}^{n-1} L^k F f^{(2k)}+L^n f^{(2n)} \tag{37}$$

$$=\sum_{k=0}^{n-1}[f^{(2k)}(0)A_k(1-t)+f^{(2k)}(1)A_k(t)]+R_n(f;\ t).\ \square$$

By means of convolution (35) we shall find a simpler expression for the remainder term $R_n(f;\ t)$.

Theorem 3.1.5. *If $n\geq 1$, then the remainder term in* (37) *has the representation*

$$R_n(f;\ t)=\frac{1}{2}\int_t^1 f^{(2n-1)}(\tau)[A_{n-1}(1-t-\tau)-A_{n-1}(1+t-\tau)]d\tau \tag{38}$$

$$+\frac{1}{2}\int_0^t f^{(2n-1)}(\tau)[A_{n-1}(1-t-\tau)+A_{n-1}(1-t+\tau)]d\tau.$$

Proof. By Theorem 3.1.3 the representation $Lf=\{t\}*f$ holds. Then

$$R_n(f;\ t)=L^n f^{(2n)}(t)=L^{n-1}(\{t\}*f^{(2n)}(t))=A_{n-1}*f^{(2n)}.$$

Now, let us simplify this expression. We have

$$-2\Lambda_{n-1}*f^{(2n)}(t)=\int_0^1\left[\int_t^0 \Lambda_{n-1}(x+t-\tau)f^{(2n)}(\tau)d\tau+\int_0^x \Lambda_{n-1}(x+t-\tau)f^{(2n)}(\tau)\,d\tau\right]dx$$

$$+\int_0^1\left[\int_{-t}^0 \Lambda_{n-1}(x-t-\tau)\,f^{(2n)}(-\tau)\,d\tau-\int_0^x \Lambda_{n-1}(x-t-\tau)f^{(2n)}(\tau)\,d\tau\right]dx,$$

where the fact that the Lidstone polynomials are odd is used. Interchanging the order of the integration in the last formula, we get

$$-2R_n(f;\,t)=-\int_0^t f^{(2n)}(\tau)\left[\int_0^1 \Lambda_{n-1}(x+t-\tau)\,dx\right]d\tau$$

$$+\int_0^1 f^{(2n)}(\tau)\left[\int_\tau^1 \Lambda_{n-1}(x+t-\tau)\,dx\right]d\tau$$

$$+\int_0^t f^{(2n)}(\tau)\left[\int_0^1 \Lambda_{n-1}(x+\tau-t)\,dx\right]d\tau-\int_0^1 f^{(2n)}(\tau)\left[\int_\tau^1 \Lambda_{n-1}(x-t-\tau)\,dx\right]d\tau.$$

Now, using the recurrency, we get

$$-2R_n(f;\,t)=\int_0^t f^{(2n)}(\tau)\,[\Lambda_n'(1-t+\tau)-\Lambda_n'(1+t-\tau)]d\tau$$

$$+\int_0^1 f^{(2n)}(\tau)[\Lambda_n'(1+t-\tau)-\Lambda_n'(1-t-\tau)]\,d\tau.$$

With elementary integration by parts and by means of routine transformations, we get

$$-2R_n(f;\,t)=\int_0^1 f^{(2n-1)}(\tau)[\Lambda_{n-1}(1+t-\tau)-\Lambda_{n-1}(1-t-\tau)]d\tau$$

$$-\int_0^t f^{(2n-1)}(\tau)[\Lambda_{n-1}(1+\tau-t)+\Lambda_{n-1}(1+t-\tau)]d\tau$$

and the representation (38) follows immediately. □

Theorem 3.1.6. *If Ψ: $\mathscr{C}([0,\,1])\to\mathbb{C}$ is a continuous linear functional, then the operation*

(39) $$(f*g)(t)=\int_0^t f(t-\tau)g(\tau)d\tau+\Psi_x\left\{\frac{\partial k(x,\,t)}{\partial x}\right\},$$

with

$$k(x,\,t)=\frac{1}{2}\int_t^x f(x+t-\tau)g(\tau)d\tau-\frac{1}{2}\int_{-t}^x f(|x-t-\tau|)g(|\tau|)\,\mathrm{sgn}\,((x-t-\tau)\tau)d\tau,$$

is a convolution of the right inverse operator of d^2/dt^2

(40) $$Lf(t)=l^2f(t)+t\Psi(f)$$

in the space $\mathscr{C}_0^1([0, 1])$ *of the smooth functions* $f(t)$ *in* $[0, 1]$ *with* $f(0)=0$.

Proof. It is not difficult to see that (39) is an inner operation for $\mathscr{C}_0^1([0, 1])$. Operator (26) is a special case of (40). The bilinearity and the commutativity of (39) are evident. The associativity of (39) can be proved in almost the same way as the associativity of (1). Here we can use at once the possibility of uniform approximation of the functions of $\mathscr{C}_0^1([0, 1])$ by odd polynomials with respect to the norm

$$\|f\|_{C^1}=\max_{0\leq t\leq 1}\{|f(t)|+|f'(t)|\}. \quad \square$$

Now we shall write down explicitly a convolution for an operator, which often finds use in problems of mathematical physics.

Example 3. Let L be the right inverse of d^2/dt^2, determined by the boundary value problem

$$y''=f,\ y(0)=0,\quad hy(1)+y'(1)=0,\quad 1+h\neq 0,$$

i. e.

$$Lf(t)=l^2f(t)+\left(\frac{h}{h+1}\int_0^1 \tau f(\tau)d\tau-\int_0^1 f(\tau)d\tau\right)t.$$

This is an operator of the form (26) with $N(f)=(hf(1)+f'(1))/(1+h)$. Then (28) gives for a convolution of L:

(41) $$(f*g)(t)=\frac{-1}{h+1}\left[k(1, t)+h\int_0^1 k(x, t)dx\right]$$

with

$$k(x, t)=\frac{1}{2}\int_t^x f(x+t-\tau)g(\tau)d\tau-\frac{1}{2}\int_{-t}^x f(|x-t-\tau|)g(|\tau|)\operatorname{sgn}((x-t-\tau)\tau)d\tau.$$

Letting $h\to\infty$, we get operation (35), as it should be expected.

3.1.3. Convolutions of the second kind right inverses of d^2/dt^2. Here we consider the class of the right inverses of d^2/dt^2 in $\mathscr{C}([0, 1])$, which are defined by a boundary value condition of the second kind and with an arbitrary linear boundary value condition, determined by a linear functional N in the space $\mathscr{C}^1([0, 1])$. Such an operator $L\colon \mathscr{C}([0, 1])\to\mathscr{C}^2([0, 1])$ is defined for each $f\in\mathscr{C}$ in such a way that $y=Lf$ to be the solution of the elementary boundary value problem

(42) $$y''=f,\quad y'(0)=0,\quad N(y)=0.$$

In order (42) to be solvable for each $f\in\mathscr{C}$, it is necessary and sufficient to assume that $N(\{1\})\neq 0$. If this condition is fulfilled, then without any loss of generality we may assume that $N(\{1\})=1$. Then

(43) $$Lf(t)=l^2f(t)-N(l^2f).$$

The defining projector F of L has the form

$$Ff(t)=f'(0)[t-N(\{x\})]+N(f).1. \tag{44}$$

Our main task is to find an explicit convolution of L if not in the whole space $\mathscr{C}([0, 1])$, then at least in some subspace of it. To this end we shall use the similarity method, reducing the operator (43) to operator of the form (40). Obviously, the transformation $T: \mathscr{C}\to\mathscr{C}_0^1([0, 1])$, defined by the Volterra integration operator

$$T: f(t)\to lf(t)=\int_0^t f(\tau)d\tau, \tag{45}$$

is a similarity of operator (43), defined in $\mathscr{C}([0, 1])$, to the operator

$$\tilde{L}\tilde{f}(t)=l^2\tilde{f}(t)-tN(l\tilde{f}) \tag{46}$$

in $\mathscr{C}_0^1([0, 1])$, i. e. $TL=\tilde{L}T$. Then (46) is an operator of the form (40) with $\Psi(f)=-N(l\tilde{f})$. Then the operation

$$(\tilde{f}\,\tilde{*}\,\tilde{g})(t)=\int_0^t \tilde{f}(t-\tau)\tilde{g}(\tau)d\tau+\Psi_x\left\{\frac{\partial}{\partial x}\left(\frac{1}{2}\int_t^x \tilde{f}(x+t-\tau)\tilde{g}(\tau)d\tau\right.\right.$$

$$\left.\left.-\frac{1}{2}\int_{-t}^x \tilde{f}(|x-t-\tau|)\tilde{g}(|\tau|)\operatorname{sgn}((x-t-\tau)\,\tau)\,d\tau\right)\right\},$$

by Theorem 3.1.6, shall be a convolution of $\tilde{L}$ in $\mathscr{C}_0^1([0, 1])$. By Theorem 1.3.6 the operation $f*g=T^{-1}[(Tf)\,\tilde{*}\,(Tg)]$ is a convolution of L in $\mathscr{C}$. Since $T^{-1}\tilde{f}=d\tilde{f}/dt$, then we get the following representation of this convolution:

$$(f*g)(t)=\int_0^t d\tau\int_0^\tau f(\tau-\sigma)\,g(\sigma)\,d\sigma+\Psi_x\left\{\frac{\partial^2\tilde{k}(x, t)}{\partial x\partial t}\right\}$$

with

$$\tilde{k}(x, t)=\frac{1}{2}\int_t^x (lf)(x+t-\tau)(lg)(\tau)d\tau$$

$$-\frac{1}{2}\int_{-t}^x (lf)(|x-t-\tau|)(lg)(|\tau|)\operatorname{sgn}((x-t-\tau)\tau)d\tau.$$

Differentiating, we get

$$\frac{\partial^2\tilde{k}}{\partial x\partial t}=\frac{1}{2}\int_t^x f(x+t-\tau)g(\tau)\,d\tau+\frac{1}{2}\int_{-t}^x (|x-t-\tau|)g(|\tau|)d\tau.$$

Thus we have proved the following

Theorem 3.1.7. *The operation*

$$(f*g)(t)=\int_0^t d\tau\int_0^\tau f(\tau-\sigma)g(\sigma)d\sigma \tag{47}$$

$$-N_x \circ l_x\left\{\frac{1}{2}\int_t^x f(x+t-\tau)g(\tau)d\tau+\frac{1}{2}\int_{-t}^x f(|x-t-\tau|)g(|\tau|)d\tau\right\}$$

is a convolution of the right inverse of d^2/dt^2, defined by

$$Lf(t)=l^2f(t)-N(l^2f),$$

where N is a linear functional on $\mathscr{C}^1([0,1])$ with $N(\{1\})=1$. *Moreover, the representation*

$$Lf=\{1\}*f \tag{48}$$

holds.

The representation (48) should be verified by a direct check.

The analogon of Theorem 3.1.6 is

Theorem 3.1.8. *If Ψ: $\mathscr{C}([0,1])\to \mathbf{C}$ is a continuous linear functional, then the operation*

$$(f*g)(t)=\int_0^t d\tau\int_0^\tau f(\tau-\sigma)g(\sigma)d\sigma+\Psi_x\left\{\frac{\partial r(x,t)}{\partial x}\right\}, \tag{49}$$

with

$$r(x,t)=\frac{1}{2}\int_t^x f(x+t-\tau)g(\tau)d\tau+\frac{1}{2}\int_{-t}^x f(|x-t-\tau|)g(|\tau|)d\tau,$$

is a convolution of the right inverse operator of d^2/dt^2, defined by

$$Lf(t)=l^2f(t)+\Psi(f) \tag{50}$$

in the space $\mathscr{C}^1([0,1])$ of the smooth functions in [0, 1], *such that the representation $Lf=\{1\}*f$ holds.*

Proof. By a direct calculation we find

$$\{\cos\lambda t\}*\{\cos\mu t\}=\frac{\cos\mu t-\cos\lambda t}{\lambda^2-\mu^2}+\frac{\mu^2E(\mu)\cos\lambda t-\lambda^2E(\lambda)\cos\mu t}{\lambda^2-\mu^2},$$

where $E(\mu)=\Psi_x\{\cos\mu x\}$, and $E(\lambda)=\Psi_x\{\cos\lambda x\}$ for arbitrary λ, μ with $\lambda\neq\mu$. We verify the associativity relation $[(\cos\lambda t)*(\cos\mu t)]*(\cos\nu t)=(\cos\lambda t)*[(\cos\mu t)*(\cos\nu t)]$ directly. Further, we differentiate this identity $2l$ times on λ, $2m$ times on μ, and $2n$ times on ν. Letting $\lambda,\mu,\nu\to 0$, we get $(\{t^{2l}\}*\{t^{2m}\})*\{t^{2n}\}=\{t^{2l}\}*(\{t^{2m}\}*\{t^{2n}\})$ for $l,m,n=0,1,2,\dots$. Hence the associativity relation $(f*g)*h=f*(g*h)$ is satisfied for even polynomials. As in the proof of Theorem 3.1.1, it is easy to show that an estimate of the form

$$\|f*g\|_1\leq A\|f\|_1\cdot\|g\|_1 \tag{51}$$

for all $f, g\in\mathscr{C}^1([0, 1])$ with $A=\text{const}$, holds. Here by $\|\ \|_1$ the integral norm $\|f\|_1=\int_0^1|f(t)|dt$ is denoted. Since the set of the even polynomials is dense in $\mathscr{L}^1([0, 1])$, then from the validity of the associativity relation in the class of the even polynomials and from the estimate (51) its validity in the whole space $\mathscr{C}^1([0, 1])$ follows. Thus the relation $(f*g)*h=f*(g*h)$ is proved as holding almost everywhere in $[0, 1]$. But since here both parts of the associativity relation are continuous functions, then it holds everywhere. Hence operation (49) is an associative operation in $\mathscr{C}^1([0, 1])$. Since $Lf=\{1\}*f$, it is a convolution of L in $\mathscr{C}^1([0, 1])$. □

Example. Let L be the right inverse of d^2/dt^2 in $\mathscr{C}([0, 1])$, defined by the solution of the boundary value problem

(52) $$y''=f,\quad y'(0)=0,\quad hy(1)+y'(1)=0\quad (h\neq 0),$$

i. e.

(53) $$Lf(t)=l^2f(t)-\int_0^1\left(1+\frac{1}{h}-\tau\right)f(\tau)d\tau.$$

This is an operator of the form (43) with $Nf=(hf(1)+f'(1))/h$. By (47) we find the convolution

(54) $$(f*g)(t)=\int_0^t d\tau\int_0^\tau f(\tau-\sigma)g(\sigma)d\sigma-\frac{1}{2h}r(1, t)-\frac{1}{2}\int_0^1 r(x, t)dx$$

with

$$r(x, t)=\int_t^x f(x+t-\tau)\,g(\tau)\,d\tau+\int_{-t}^x f(|x-t-\tau|)g(|\tau|)d\tau.$$

The boundary value problem (52) is often used in problems of mathematical physics.

3.1.4. Convolutions of the third kind right inverses of d^2/dt^2. Now let L be a right inverse operator of d^2/dt^2 in $\mathscr{C}([0, 1])$, defined by a boundary value condition of the third kind and by a general boundary value condition, determined by an arbitrary linear functional N in $\mathscr{C}^1([0, 1])$. If $f\in\mathscr{C}^1([0, 1])$, we define $y=Lf$ as the solution of the boundary value problem

(55) $$y''=f,\quad y'(0)-hy(0)=0,\quad N(y)=0,$$

where h is a given constant, and N is an arbitrary non-zero linear functional on $\mathscr{C}^1([0, 1])$. For the solvability of (55) for each $f\in\mathscr{C}([0, 1])$ it is necessary and sufficient the condition $N\{1+hx\}\neq 0$ to be satisfied. It is clear that without any loss of generality we may assume that

(56) $$N\{1+hx\}=1.$$

Then the solution $y=Lf$ of (56) is

(57) $$Lf(t)=l^2f(t)-N(l^2f)(1+ht).$$

The defining projector F of L has the form

(58) $$Ff(t)=[t-N\{x\}](f'(0)-hf(0))+(1+ht)N(f).$$

Since the boundary value problem (42) is a special case of (57), we shall try to proceed in a similar way. Here the approach of Dimovski and Petrova [50] is presented. First, we find a similarity from (57) to an operator of the form (43).

Lemma 3. *The linear transformation*

(59) $$Tf(t)=f(t)-h\int_0^t e^{-h(t-\tau)}f(\tau)d\tau$$

is an isomorphism of the linear space $\mathscr{C}([0,1])$ *on itself, and its inverse is the transformation*

(60) $$T^{-1}f(t)=f(t)+h\int_0^t f(\tau)d\tau.$$

Proof. By a direct check. □

Lemma 4. *Transformation* (59) *is a similarity of the right inverse* (57) *of* d^2/dt^2 *to the right inverse*

(61) $$\tilde{L}f(t)=l^2f(t)-\tilde{N}(l^2f)$$

of d^2/dt^2 *with* $\tilde{N}=N\circ T^{-1}$, *i. e.* $TL=\tilde{L}T$.

Proof. It is obvious that T commutes with l: $Tl=lT$ and that $T\{1+ht\}=1$. Then

$$TLf=Tl^2f-N(l^2f)T(1+ht)=l^2Tf-(N\circ T^{-1})(l^2Tf)$$

and the assertion is proved. □

Theorem 3.1.9. *The operation*

(62) $$(f*g)(t)=N\{1\}(f\overset{h}{*}g)(t)-(N\circ l)_x\{-h(f\overset{h}{*}g)(|x-t|)+\frac{1}{2}\int_t^x f(x+t-\tau)g(\tau)d\tau+\frac{1}{2}\int_{-t}^x f(|x-t-\tau|)g(|\tau|)d\tau\},$$

with

(63) $$(f\overset{h}{*}g)(t)=\int_0^t e^{-h(t-\tau)}\left(\int_0^\tau f(\tau-\sigma)g(\sigma)d\sigma\right)d\tau,$$

is a convolution of the right inverse (57) *of* d^2/dt^2 *in* $\mathscr{C}([0,1])$, *such that the representation*

(64) $$Lf=\{1+ht\}*f$$

holds.

Proof. By Lemmas 3 and 4 and by Theorem 3.1.6, the operation

$$(f \mathbin{\widehat{*}} g)(t) = T^{-1}[(Tf) \mathbin{\tilde{*}} (Tg)], \tag{65}$$

where $\tilde{*}$ is operation (47), but with $\tilde{N}$ instead of N, is a convolution of (57) in $\mathscr{C}([0, 1])$. This is an immediate implication of the general Theorem 1.5.6. Moreover, $Lf = \{1+ht\} * f$, since $T\{1+ht\} = 1$. We shall show that convolution (65) coincides with (62). The identity

$$(f * g)(t) = (f \mathbin{\widehat{*}} g)(t) \tag{66}$$

can easily be verified for $f(t) = f_\lambda(t) = \cos \lambda t + (h/\lambda) \sin \lambda t$ and $g(t) = f_\mu(t) = \cos \mu t + (h/\mu) \sin \mu t$ with arbitrary real λ, $\mu \neq 0$. Using the obvious identities $f_\lambda(t) = T^{-1}\{\cos \lambda t\}$ and $f_\mu(t) = T^{-1}\{\cos \mu t\}$, the desired identity (66) can be written in the form

$$(T^{-1}\{\cos \lambda t\}) * (T^{-1}\{\cos \mu t\}) = (T^{-1}\{\cos \lambda t\}) \mathbin{\widehat{*}} \{T^{-1}(\cos \mu t)\}. \tag{67}$$

The last identity is satisfied also in the case when λ or μ is 0. Since $T^{-1}\{\cos 0 . t\} = 1+ht$, then by $f_0(t)$ we can understand the function $1+ht$. We need the identity (67) for integer multiples of π only, i. e. for λ, $\mu = 0$, π, $2\pi, \ldots$. The system $\{\cos n\pi t\}_{n=0}^{\infty}$ satisfies the hypothesis of the Weierstrass-Stone approximation theorem. Taking into account the continuity of the inverse transformation T^{-1}, along with the continuity of the operations $*$ and $\widehat{*}$ with respect to the uniform convergence in $\mathscr{C}([0, 1])$, we conclude at once that (66) is fulfilled for arbitrary f, $g \in \mathscr{C}([0, 1])$. The theorem is proved. □

Example 1. Let $h=0$. Then $N\{1\}=1$, due to the assumption $N\{1+hx\}=1$. Then convolution (62) takes the form

$$(f * g)(t) = \int_0^t d\tau \int_0^\tau f(\tau-\sigma) g(\sigma) d\sigma$$

$$-(N \circ l)_x \left\{ \frac{1}{2} \int_t^x f(x+t-\tau) g(\tau) d\tau + \frac{1}{2} \int_{-t}^x f(|x-t-\tau|) g(|\tau|) d\tau \right\},$$

i. e. it coincides with (47), as it should be expected.

Example 2. Let χ be a linear functional in $\mathscr{C}^1([0, 1])$, with $\chi(\{1\})=0$ and $\chi(\{x\})=1$. Then, let us take the functional $N=\chi/h$. It satisfies the condition $N(\{1+hx\})=1$ and hence operation (62), which we shall denote by $(f * g)_h$, is a convolution of the operator

$$L_h f(t) = l^2 f(t) - \chi(l^2 f)(1+ht)/h$$

such that $L_h f = [(1+ht) * f]_h$. Letting $h \to +\infty$, we see that $\lim\limits_{h\to+\infty} (f * g)_h = 0$, but the expression $h(f * g)_h$ has a finite limit, which we shall denote by $f * g$. Then

$$(f * g)(t) = -\chi_x \left\{ -\int_0^{|x-t|} f(|x-t|-\tau) g(\tau) d\tau \right.$$

$$+\frac{1}{2}\int_0^x f(x+t-\tau)g(\tau)d\tau+\frac{1}{2}\int_{-t}^x f(|x-t-\tau|)g(|\tau|)d\tau\Big\}.$$

This operation is a convolution of the operator $Lf(t)=l^2f(t)-\chi(lf).t$ with $Lf=\{t\}*f$. This is exactly a convolution of the type (28), provided we take into account (18) and Lemma 1.

Example 3. Let L be the right inverse operator of d^2/dt^2 defined by the boundary value problem

$$y''=f,\quad y'(0)-hy(0)=0,\quad y'(1)+Hy(1)=0. \tag{68}$$

In order this boundary value problem to be solvable for each $f\in\mathscr{C}([0,1])$, we should assume $h+H+hH\neq 0$. Since $N(f)=[f'(1)+Hf(1)]/(h+H+hH)$, then

$$(N\circ l)f=\frac{1}{h+H+hH}\left[f(1)+H\int_0^1 f(\tau)\,d\tau\right].$$

Convolution (62) in this case takes the form

$$(f*g)(t)=\frac{1}{h+H+hH}\left[H(f\overset{h}{*}g)(t)-A_h(1,t)-\int_0^1 A_h(x,t)\,dx\right], \tag{69}$$

where

$$A_h(x,t)=-h(f\overset{h}{*}g)(|x-t|)+\frac{1}{2}\int_t^x f(x+t-\tau)g(\tau)d\tau$$

$$+\frac{1}{2}\int_{-t}^x f(|x-t-\tau|)g(|\tau|)d\tau$$

and where by $f\overset{h}{*}g$ the operation

$$(f\overset{h}{*}g)(t)=\int_0^t e^{-h(t-\tau)}\left[\int_0^\tau f(\tau-\sigma)g(\sigma)d\sigma\right]d\tau$$

is denoted.

3.1.5. Operational calculi for right inverses of the square of differentiation. The convolutions found in the previous sections can be used for developing operational calculi, intended for some classes of boundary value problems. In this we can follow the general scheme of Chapter 1. Here two approaches could be used: the convolution quotients approach and that of the multiplier quotients. Though they are equivalent, the latter approach has some advantages in applications to boundary value problems of mathematical physics. Here we confine ourselves to the multiplier quotients approach, though the interpretation of the results obtained as convolution quotients is immediate. Moreover, we consider the convolutions introduced in the previous sections only.

1. Multiplier quotients rings for first kind right inverses of the square of differentiation.

As first example, we shall consider the right inverse operator

$$Lf(t)=l^2f(t)-t\int_0^1(1-\tau)f(\tau)d\tau \tag{70}$$

of d^2/dt^2, defined as the solution of the two-point boundary value problem (32). As we had seen in 3.1.2, the operation

$$(f*g)(t)=-\int_0^1 k(x,t)dx \tag{71}$$

with

$$k(x,t)=\frac{1}{2}\int_t^x f(x+t-\tau)g(\tau)d\tau-\frac{1}{2}\int_{-t}^x f(|x-t-\tau|)g(|\tau|)\operatorname{sgn}((x-t-\tau)\tau)d\tau,$$

is a convolution of L in $\mathscr{C}([0,1])$, such that $Lf=\{t\}*f$. The expression

$$Ff(t)=f(0)(1-t)+f(1)t \tag{72}$$

represents the defining projector of (70) in the whole space $\mathscr{C}([0,1])$.

As first task, let us characterize the divisors of zero of (71). To this end we shall find the eigenvalues and eigenfunctions of L. It is easy to see that all the eigenvalues of L are $\lambda_n=-1/n^2\pi^2$, $n=1,2,\ldots$. The corresponding eigenfunctions are $\varphi_n(t)=\sin n\pi t$, $n=1,2,\ldots$. If $f(t)$ is an arbitrary function of $\mathscr{C}([0,1])$, then it can easily be found that

$$(f*\varphi_n)(t)=\chi_n(f)\varphi_n(t),\quad n=1,2,\ldots, \tag{73}$$

with

$$\chi_n(f)=\frac{(-1)^n}{n\pi}\int_0^1 f(t)\sin n\pi t\,dt.$$

Theorem 3.1.10. *A function $f\in\mathscr{C}([0,1])$ is a non-trivial non-divisor of zero of* (71) *iff*

$$\int_0^1 f(t)\sin n\pi t\,dt\neq 0 \tag{74}$$

for $n=1,2,\ldots$.

Proof. The proof follows immediately from the multiplicativity property

$$\chi_n(f*g)=\chi_n(f)\chi_n(g) \tag{75}$$

of the functionals χ_n and from the well-known fact that the system $\{\sin n\pi t\}$, $n=1,2,\ldots$, is total in $\mathscr{C}([0,1])$ in the sense that $\chi_n(f)=0$ for $n=1,2,\ldots$, imply $f\equiv 0$. Let f be a divisor of zero of (71), i.e. let there exist a function

$g\in\mathscr{C}([0,1])$, $g\not\equiv 0$, such that $f*g\equiv 0$. From (75) it follows that $\chi_n(f)\cdot\chi_n(g)=0$ for $n=1, 2, \dots$. Since $g\not\equiv 0$, there exists at least one k, such that $\chi_k(g)\neq 0$. Then $\chi_k(f)=0$ and thus (74) is violated. Conversely, if for some $m>0$ we have $\int_0^1 f(t)\sin m\pi t dt=0$, then from (73) it follows that $f*\{\sin m\pi x\}=0$ and hence f is a divisor of zero, provided it is not identically zero. □

Corollary 1. A non-zero function $f\in\mathscr{C}([0,1])$ is a divisor of zero of (71) iff at least one of the Fourier sine-coefficients $\int_0^1 f(t)\sin n\pi t dt$ of f is equal to zero.

Corollary 2. A convolutional operator

$$Mf=m*f, \tag{76}$$

with $m\in\mathscr{C}([0,1])$, and with (71) as the convolution $*$, is a divisor of zero in the multipliers ring $\mathfrak{M}$ of (71), iff m is a divisor of zero of convolution (71).

Let M be the multiplier quotients ring of (71). We shall denote the convolutional multipliers $Mf=m*f$ by $M=m*$. In particular, the representation $Lf=\{t\}*f$ means that $L=\{t\}*$. In other words, the convolutional operators ring is isomorphic with the ring of the continuous functions $\mathscr{C}([0,1])$ with the multiplication (71). From now on the inverse element of L in M shall be denoted by S, i. e. $S=1/L$. In applications of the corresponding operational calculus, explicit representations of the basic partial fraction with S are needed.

Theorem 3.1.11. *If λ is an arbitrary complex number, such that $\lambda\neq -k^2\pi^2$, $k=1, 2, \dots$, then for each integer $n\neq 0$ the rèpresentation*

$$\frac{1}{(S+\lambda)^n}=\left\{\frac{(-1)^{n-1}}{(n-1)!}\frac{\partial^{n-1}}{\partial\lambda^{n-1}}\left(\frac{\sin\sqrt{\lambda}\,t}{\sin\sqrt{\lambda}}\right)\right\}* \tag{77}$$

holds. In particular,

$$\frac{1}{S+\lambda}=\left\{\frac{\sin\sqrt{\lambda}\,t}{\sin\sqrt{\lambda}}\right\}*,$$

where for $\sqrt{\lambda}$ one and the same value is taken.

Proof. Let us prove (77) in the case $n=1$. Denote $\varphi=\{\sin\sqrt{\lambda}t/\sin\sqrt{\lambda}\}$. Since $1/(S+\lambda)=L/(1+\lambda L)$, then the desired formula takes the form $L=(1+\lambda L)\varphi *$. It is equivalent to $L=[(1+\lambda L)\varphi]*$, i. e. to $\{t\}*=[(1+\lambda L)\varphi]*$. Therefore, we should prove that the equality $\{t\}=(1+\lambda L)\varphi$ is an identity. But this is a matter of an elementary check.

The proof of (77) for arbitrary $n>0$ proceeds as the proof of a similar representation in 2.3.1.

The example considered is a pattern for construction of multiplier quotients rings for all first kind right inverses of the square of differentiation. Indeed, if we take the general right inverse operator

$$Lf(t)=l^2f(t)-N(l^2f)t \tag{78}$$

of d^2/dt^2, considered in 3.1.2, with an arbitrary linear functional N in $\mathscr{C}^1([0, 1])$ with $N\{x\}=1$, then, by Theorem 3.1.3, the operation

(79)
$$(f*g)(t)=-\frac{1}{2}\Phi_x\Big\{\int_t^x f(x+t-\tau)g(\tau)\,d\tau$$
$$-\int_{-t}^x f(|x-t-\tau|)g(|\tau|)\,\mathrm{sgn}\,((x-t-\tau)\tau)d\tau\Big\},$$

with $\Phi=N\circ l$, is a convolution of L in $\mathscr{C}([0, 1])$. Here the representation $Lf=\{t\}*f$ holds too. In contrast to the example considered, in the general case we are not able to give an exhaustive characterization of the divisors of zero of the convolution (79). However, for the purposes of operational calculus this is not obligatory. In finding representations of partial fractions of the form $1/(S+\lambda)^n$, it is necessary only to know whether $S+\lambda$ is non-divisor of zero. But from the general theory of 1.4.4 we know that $S+\lambda$ is a divisor of zero in M iff $-1/\lambda$ is an eigenvalue of L. Hence $S+\lambda$ is not a divisor of zero in M iff λ is not an eigenvalue of the elementary boundary value problem

$$y''+\lambda y=0,\ y(0)=0,\ N(y)=0.$$

It is obvious that the eigenvalues of this problem are the zeros of the entire function $E(\lambda)=\frac{1}{\sqrt{\lambda}}N_x\{\sin x\sqrt{\lambda}\}$. Then, as in the proof of Theorem 3.1.11, it is easily seen that if λ is not a zero of $E(\lambda)$, i. e. if $E(\lambda)\neq 0$, then the representation

(80)
$$\frac{1}{(S+\lambda)^n}=\frac{(-1)^{n-1}}{(n-1)!}\frac{\partial^{n-1}}{\partial\lambda^{n-1}}\left\{\frac{\sin\sqrt{\lambda}\,t}{\sqrt{\lambda}\,E(\lambda)}\right\}*$$

holds.

2. *Multiplier quotients rings for second kind right inverses of the square of differentiation.*

First we shall consider the right inverse operator L of d^2/dt^2 in $\mathscr{C}([0, 1])$ defined by means of the boundary value problem

$$y''=f,\quad y'(0)=0,\quad hy(1)+y'(1)=0,\quad h\neq 0,$$

i. e. the operator

(81)
$$Lf(t)=l^2 f(t)+\int_0^1 \Big(1+\frac{1}{h}-\tau\Big)f(\tau)\,d\tau.$$

In 3.1.3 we had found as a convolution of L in $\mathscr{C}([0, 1])$ the operation

(82)
$$(f*g)(t)=\int_0^t d\tau\int_0^\tau f(\tau-\sigma)g(\sigma)d\sigma-\frac{1}{2h}r(1,\ t)-\frac{1}{2}\int_0^1 r(x,\ t)dx$$

with

$$r(x,\ t)=\int_t^x f(x+t-\tau)\,g(\tau)d\tau+\int_{-t}^x f(|x-t-\tau|)g(|\tau|)d\tau.$$

Then the representation $Lf=\{1\}*f$ holds. The eigenvalues of the problem

(83) $$y''+\lambda y=0,\quad y'(0)=0,\quad hy(1)+y'(1)=0$$

are the zeros of the entire function $E(\lambda)=[h\cos\sqrt{\lambda}-\sqrt{\lambda}\sin\sqrt{\lambda}]/h$ If $\lambda\in C$ is such that $E(\lambda)\neq 0$, then we get

(84) $$\frac{1}{(S+\lambda)^n}=\left\{\frac{(-1)^{n-1}}{(n-1)!}\frac{\partial^{n-1}}{\partial\lambda^{n-1}}\left(\frac{\cos\sqrt{\lambda}\,t}{E(\lambda)}\right)\right\}*.$$

We shall not consider the general case, since everything proceeds in the same way as in the previous subsection.

3. Multiplier quotients rings for third kind right inverses of the square of differentiation.

In 3.1.4 we have found a convolution for the general right inverse of d^2/dt^2 of the third kind. For a corresponding operational calculus we shall confine ourselves to the right inverse considered in Example 3 of this section. This right inverse L of d^2/dt^2 is defined by the elementary boundary value problem

$$y''=f,\ y'(0)-hy(0)=0,\ y'(1)+Hy(1)=0,$$

under the assumption that $h+H+hH\neq 0$. It has the form

(85) $$Lf(t)=l^2f(t)-\frac{(1+ht)}{h+H+hH}\left[\int_0^1 f(\tau)d\tau+H\int_0^1(1-\tau)f(\tau)d\tau\right].$$

The convolution of L in $\mathscr{C}([0,1])$, found in 3.1.4, had the form (69). The exact form of this convolution is not necessary for general considerations and for finding representations of the partial fractions $1/(S+\lambda)^n$ as convolutional operators in this convolution. It is enough to know the representation $Lf=\{1+ht\}*f$ only. The element $S+\lambda$ of M is invertible iff λ is not a zero of the entire function $E(\lambda)=(h+H)\cos\sqrt{\lambda}+(hH-\lambda)\sin\sqrt{\lambda}/\sqrt{\lambda}$. In such a case the representation

(86) $$\frac{1}{(S+\lambda)^n}=\left\{\frac{(-1)^{n+1}}{(n-1)!}\frac{\partial^{n-1}}{\partial\lambda^{n-1}}\left(\frac{\cos\sqrt{\lambda}\,t+h\sin\sqrt{\lambda}\,t/\sqrt{\lambda}}{(h+H+hH)E(\lambda)}\right)\right\}*$$

holds. In particular

$$\frac{1}{S+\lambda}=\left\{\frac{\cos\sqrt{\lambda}\,t+h\sin\sqrt{\lambda}\,t/\sqrt{\lambda}}{(h+H+hH)E(\lambda)}\right\}*.$$

3.2. CONVOLUTIONS OF INITIAL VALUE RIGHT INVERSFS OF LINEAR SECOND-ORDER DIFFERENTIAL OPERATORS

In this section we consider the problem of determining convolutions of right inverse operators of linear differential operators of the second order with variable coefficients, defined by initial value problems for a point of the interval considered. First we treat the case of non-singular linear differential operators of the second order, and then as an example for a singular second-order linear differential operator we take the Bessel differential operator. In both cases the basic method for finding convolutions of the corresponding right inverses is the similarity method, developed in general terms in 1.3.3. The Delsarte-Povzner transmutation operators and the Sonine transform are used as similarities.

3.2.1. Convolutions of the initial value right inverse of non-singular second-order linear differential operator. In this subsection we consider the general linear differential operator of the second order

(1) $$D=a(x)\frac{d^2}{dx^2}+b(x)\frac{d}{dx}+c(x)$$

in an interval Δ, assuming that the coefficients are complex-valued and satisfy the smoothness conditions $a(x)\in\mathscr{C}^2(\Delta)$, $b(x)\in\mathscr{C}^1(\Delta)$ and $c(x)\in\mathscr{C}(\Delta)$. Moreover, we assume that $a(x)$ is real-valued, and $a(x)>0$. Let x_0 be an arbitrary but fixed point of Δ.

Definition 1. A linear operator L_0: $\mathscr{C}(\Delta)\to\mathscr{C}^2(\Delta)$ is said to be the initial value right inverse of D for the point $x=x_0$, if for each $f\in\mathscr{C}(\Delta)$ the function $y=Lf$ is the solution of the initial value problem

(2) $$Dy=f,\quad y(x_0)=y'(x_0)=0.$$

In order to express explicitly the defining projector F of L, we introduce the fundamental system $y_1(x)$, $y_2(x)$ of the operator D for the point x_0. These are the solutions of $Dy=0$, which satisfy the initial value conditions $y_1(x_0)=1$, $y_1'(x_0)=0$, $y_2(x_0)=0$, $y_2'(x_0)=1$. Then for $Ff=(I-LD)f$ we get

(3) $$Ff(x)=f(x_0)y_1(x)+f'(x_0)y_2(x)$$

and it can be considered on $\mathscr{C}^1(\Delta)$ (Bozhinov, Dimovski [17]). We shall show that L_0 has a convolution in $\mathscr{C}(\Delta)$. Without an essential loss of generality, we may consider the operator

(4) $$D=\frac{d^2}{dt^2}-q(t)$$

instead of (1).

Lemma 1. *The transformation T: $\mathscr{C}(\Delta)\to\mathscr{C}(\Delta_0)$ given by*

(5) $$T: f(x)\to\varrho(t)f[\varphi^{-1}(t)],$$

with

$$\varphi(x)=\int_{x_0}^{x}\frac{d\xi}{\sqrt{a(\xi)}}\quad\text{and}\quad\varrho(t)=\exp\left\{\int_{x_0}^{\varphi^{-1}(t)}\frac{2b(\xi)-a'(\xi)}{4a(\xi)}\,d\xi\right\},$$

where $\Delta_0=\varphi(\Delta)$, transforms $a(x)(d^2/dx^2)+b(x)(d/dx)+c(x)$ into $(d^2/dt^2)-q(t)$, where $q(t)=\left[-c(x)+\frac{4b^2(x)-a'(x)^2}{16a(x)}+a(x)\frac{d}{dx}\left(\frac{2b(x)-a'(x)}{4a(x)}\right)\right]_{x=\varphi^{-1}(t)}$ in the sense of the similarity

(6) $$T\left[a(x)\frac{d^2}{dx^2}+b(x)\frac{d}{dx}+c(x)\right]=\left[\frac{d^2}{dt^2}-q(t)\right]T.$$

The proof consists in an elementary verification and in essence it is contained in every textbook on ordinary differential equations.

Further we consider operator (4) only, denoting it by D, and let the interval in which t varies be Δ. But the fundamental system $u_1(t)$, $u_2(t)$ shall be considered for the point $t=0$. Under these assumptions the defining projector F of L_0 has the form

$$Ff=f(0)u_1(t)+f'(0)u_2(t). \tag{7}$$

Now we shall show that there exists a similarity T of L_0 to l^2, which is a Volterra transformation of the second kind

$$Tf(t)=f(t)+\int_0^t K(t,\tau)f(\tau)d\tau. \tag{8}$$

Lemma 2. *Let $q(t)\in\mathscr{C}^1(\Delta)$, and let $K(t,\tau)$ be a solution of the partial differential equation*

$$K_{tt}=K_{\tau\tau}-q(\tau)K \tag{9}$$

in the domain $G=\{(t,\tau):t\in\Delta,\min(t,0)\leq\tau\leq\max(t,0)\}$, which satisfies the condition

$$K(t,t)=-\frac{1}{2}\int_0^t q(\tau)d\tau. \tag{10}$$

If T is transformation (8) *with this $K(t,\tau)$ as kernel, then for every $f(t)\in\mathscr{C}^2(\Delta)$, the identity*

$$\left(TD-\frac{d^2}{dt^2}T\right)f(t)=f(0)K_\tau(t,0)-f'(0)K(t,0) \tag{11}$$

holds,

Proof. Obviously, T is a linear isomorphism of $\mathscr{C}(\Delta)$ onto itself, i. e. an automorphism, and $\mathscr{C}^1(\Delta)$ and $\mathscr{C}^2(\Delta)$ are invariant subspaces of T. It is clear that T preserves the values of the functions and of their derivatives in $t=0$, i. e. $Tf(0)=f(0)$ and $(Tf)'(0)=f'(0)$. The assertion of the lemma follows from the easily verified identity

$$\left(TD-\frac{d^2}{dt^2}T\right)f(t)=\int_0^t\left[K_{\tau\tau}(t,\tau)-q(\tau)K(t,\tau)-K_{tt}(t,\tau)\right]f(\tau)d\tau \tag{12}$$

$$-\left[2\frac{d}{dt}K(t,t)+q(t)\right]f(t)+f(0)K_\tau(t,0)-f'(0)K(t,0). \quad\square$$

It is not difficult to show the existence of kernels which satisfy (9) and the boundary value condition (10). To see this, let us make the characteristic change of the variables

$$u=(t+\tau)/2,\quad v=(t-\tau)/2. \tag{13}$$

Then for $k(u, v)=K(u+v, u-v)$ we get the equation

$$k_{uv}=-q(u-v)k \tag{14}$$

in the corresponding domain G'. If along with (10) we impose the condition

$$K(t, 0)=\alpha(t) \tag{15}$$

with an arbitrarily given function $\alpha(t)\in\mathscr{C}(\Delta)$, then for $k(u, v)$ we get the conditions

$$k(u, 0)=-\frac{1}{2}\int_0^u q(\xi)d\xi, \quad k(u, u)=\alpha(u). \tag{16}$$

It is easy to see that the boundary value problem (14), (16) can be reduced to the second kind Volterra integral equation

$$k(u, v)=\alpha(v)-\frac{1}{2}\int_v^u q(\xi)d\xi-\int_v^u d\xi\int_0^v q(\xi-\eta)k(\xi, \eta)d\eta. \tag{17}$$

From the well-known theory of the Volterra integral equations (see, e. g Marchenko [74], p. 19) follow the existence and the uniqueness of a solution of (17) in the whole domain G. This solution could be found e. g. by the successive approximations method. It is essential to note that (17) has a solution for each $q\in\mathscr{C}(\Delta)$, but it may happen $k(u, v)$ to be only once differentiable, if $q(t)$ is not differentiable. Nevertheless, if T is (8) with the kernel

$$K(t, \tau)=k\left(\frac{t+\tau}{2}, \frac{t-\tau}{2}\right), \tag{18}$$

then it can be shown that $T: \mathscr{C}^2(\Delta)\to\mathscr{C}^2(\Delta)$ and (11) is also valid. Since we do not use this property, we shall confine ourselves to a weaker proposition.

Lemma 3. *If* $q(t)\in\mathscr{C}(\Delta)$, *and* $K(t, \tau)$ *is* (18), *then* $Tf(t)=f(t)+\int_0^t K(t, \tau)f(\tau)d\tau$ *is a similarity of* L_0 *to* l^2 *in* $\mathscr{C}(\Delta)$, *i. e. the relation*

$$TL_0=l^2T \tag{19}$$

holds.

Proof. If $q\in\mathscr{C}^1(\Delta)$, then (19) follows immediately from identity (11). Indeed if we substitute $L_0f(t)$ instead of $f(t)$, for $f(t)\in\mathscr{C}(\Delta)$ we get

$$TDL_0f(t)-\frac{d^2}{dt^2}TL_0f(t)=(L_0f)(0)K_\tau(t, 0)-(L_0f)'(0)K(t, 0),$$

i. e. $(d^2/dt^2)[TL_0f(t)]=Tf(t)$. Since $L_0f(0)=(L_0f)'(0)=0$ then $TL_0f(t)=l^2Tf(t)$.

Now, let us consider the general case $q(t)\in\mathscr{C}(\Delta)$. In order to prove (19) in the class $\mathscr{C}(\Delta)$, we approximate $q(t)$ by a sequence $q_n(t)\in\mathscr{C}^1(\Delta)$, $n=1, 2, \ldots$, in the topology of the almost uniform convergence in $\mathscr{C}(\Delta)$. Let $\tilde{\Delta}$ with

$\tilde{\Delta}\subset\Delta$ be a compact subinterval of Δ, containing the point $t=0$. Then the sequence $\{q_n(t)\}_{n=1}^{\infty}$ is iniformly convergent to $q(t)$ in $\tilde{\Delta}$. Since $\mathscr{C}(\tilde{\Delta})$ is a Banach space, then using the usual methods of the theory of Banach spaces, we can easily see that the operator sequences $\{T_n\}_{n=1}^{\infty}$ and $\{L_{0,n}\}_{n=1}^{\infty}$, which correspond to the functions of the sequence $\{q_n\}_{n=1}^{\infty}$, are convergent to the restrictions of T and L_0 to $\mathscr{C}(\tilde{\Delta})$. Then, in order to establish (19) in $\mathscr{C}(\Delta)$, we let $n\to\infty$ in $T_nL_{0,n}f=l^2T_nf$. □

Later on, for definiteness, we suppose that the arbitrary function $\alpha(t)$ in (17) is chosen to be 0.

Theorem 3.2.1. *The operation*

$$f\overset{\circ}{*}g=T^{-1}[(Tf)*(Tg)], \tag{20}$$

where $*$ denotes the Duhamel convolution in $\mathscr{C}(\Delta)$, is a continuous convolution of the initial value right inverse operator L_0 of $D=(d^2/dt^2)-q(t)$ in $\mathscr{C}(\Delta)$, and the representation $Lf=u_2\overset{\circ}{*}f$ holds.

The proof follows immediately from Theorem 1.3.6. The continuity of (20) is ensured by the continuity of the Duhamel convolution and by the continuity of the transformations T and T^{-1}. The representation $Lf=u_2\overset{\circ}{*}f$ follows directly from the representation $l^2f=\{t\}*f$ and from $u_2=T^{-1}\{t\}$. □

Theorem 3.2.2. *The convolution quotients ring of $\mathscr{C}(\Delta)$ with respect to convolution* (20) *is isomorphic to the Mikusiński ring for $\mathscr{C}(\Delta)$.*

Proof. It is easy to see that the similarity T is an algebra isomorphism of the convolutional algebras $(\mathscr{C}(\Delta),\overset{\circ}{*})$ and $(\mathscr{C}(\Delta),*)$. Therefore, from the results of 1.4.3, their convolution quotients rings are isomorphic. □

Hence, there are two ways for developing operational calculus for the operator L_0. One of them is to proceed directly, using convolution (20) and finding directly the corresponding partial fractions. The other approach consists in transferring all the facts of the classical operational calculus to facts of the operational calculus we are to develop. By means of convolution (20) a representation of all continuous linear operators $M:\mathscr{C}(\Delta)\to\mathscr{C}(\Delta)$, with an invariant subspace

$$\mathscr{C}_0^2(\Delta)=\{f\in\mathscr{C}^2(\Delta),\ f(0)=f'(0)=0\}, \tag{21}$$

which commute with $D=(d^2/dt^2)-q(t)$ in $\mathscr{C}_0^2(\Delta)$, can be found (for details see Bozhinov, Dimovski [20]). Here for the sake of simplicity we confine ourselves to the simpler case of the one-sided interval $\Delta=[0,\infty)$.

Theorem 3.2.3. *A continuous linear operator $M:\mathscr{C}([0,\infty))\to\mathscr{C}([0,\infty))$, which maps $\mathscr{C}_0^2([0,\infty))$ into itself, commutes with $D=(d^2/dt^2)-q(t)$ in $\mathscr{C}_0^2([0,\infty))$, iff it admits a representation of the form*

$$Mf=D(m\overset{\circ}{*}f), \tag{22}$$

where $m\in\mathscr{C}^1([0,\infty))$, and m' is a function with locally bounded variation in $[0,\infty)$.

Proof. It is easy to see that the problem for finding the operators M which satisfy the hypothesis of the theorem is equivalent with the problem for finding the commutant of L_0 in $\mathscr{C}([0, \infty))$. By means of the similarity T, the last problem is reduced to the problem of finding the commutant of l^2 in $\mathscr{C}([0, \infty))$. In 1.1.3 it was shown that a linear continuous operator $\tilde{M}: \mathscr{C}(\Delta)\rightarrow\mathscr{C}(\Delta)$ commutes with the Volterra integration operator l in $\mathscr{C}(\Delta)$ iff it admits a representation of the form

(23) $$\tilde{M}f=\frac{d}{dt}(\tilde{m}*f)=\frac{d^2}{dt^2}[l\tilde{m}*f],$$

where $\tilde{m}=\tilde{M}\{1\}$ is a function with a locally bounded variation in $[0, \infty)$. Since in the case of the one-sided interval the commutants of l and l^2 coincide (see Bozhinov, Dimovski [20]), it is clear that (23) represents the commutant of l^2 in $\mathscr{C}([0, \infty))$. Let $M: \mathscr{C}([0, \infty))\rightarrow\mathscr{C}([0, \infty))$ be an arbitrary continuous operator, which commutes with L_0 in $\mathscr{C}([0, \infty))$. Let us consider the continuous linear operator $\tilde{M}=TMT^{-1}$, which obviously commutes with l^2 in $\mathscr{C}([0, \infty))$. Therefore a representation of the form (23) holds for M. Since $(l\tilde{m}*Tf)(0)=(l\tilde{m}*Tf)'(0)=0$, then from formula (11) we get

$$Mf = T^{-1}\tilde{M}T\,f=T^{-1}\,(d^2/dt^2)[l\,\tilde{m}*Tf]=DT^{-1}[(Tm)*(Tf)]=D[m\overset{\circ}{*}f],$$

where we have denoted $m=T^{-1}(l\tilde{m})$. Obviously, $m\in\mathscr{C}^1([0, \infty))$. It is not so obvious that m' is a function with locally bounded variation, when $q\in\mathscr{C}([0, \infty))$. Then the kernel $K(t, \tau)$ is not from the $\mathscr{C}^2$-class. But this assertion follows easily from the identity

(24) $$(Tf)'=f'(t)+K(t, t)f(t)+\int_0^t K_t(t, \tau)f(\tau)d\tau.$$

Since $\alpha(t)$ is taken to be 0, from integral equation (17) it can be seen that

(25) $$K_t(t, \tau)=-\frac{1}{4}q\left(\frac{t+\tau}{2}\right)+\frac{1}{4}q\left(\frac{t-\tau}{2}\right)+F(t, \tau),$$

where $F(t, \tau)$ is a smooth function from $\mathscr{C}^1(G)$. But it can be shown directly by a linear substitution that the functions

$$\int_0^t q\left(\frac{t+\tau}{2}\right)f(\tau)d\tau,\quad \int_0^t q\left(\frac{t-\tau}{2}\right)f(\tau)d\tau$$

are smooth ones, provided $f\in\mathscr{C}^1([0, \infty))$. If f' is a function with locally bounded variation, then the function Tf is a smooth one and $(Tf)'$ is a function with locally bounded variation, and vice versa. □

In the case considered, when $\Delta=[0, \infty)$, an integral transformation can be defined, which is to play the same role as the role played by the Laplace transform in the classical operational calculus. Let $\mathscr{C}_{\exp}$ denote the subspace of $\mathscr{C}([0, \infty))$, consisting of the functions which are $O(e^{\alpha t})$ for $t\rightarrow+\infty$, and let $\mathscr{C}_T=T^{-1}(\mathscr{C}_{\exp})$. Then we define the integral transform

(26) $$\mathfrak{R}\{f;\, p\}=\int_0^\infty e^{-pt}(Tf)(t)dt$$

as the composition of the Laplace transform $\mathfrak{L}\{f;\, p\}=\int\limits_0^\infty e^{-pt}f(t)dt$ and the similarity T.

Theorem 3.2.4. *For arbitrary $f,\ g\in\mathscr{C}_T$ the identities*

$$\mathfrak{R}\{L_0f;\, p\}=\frac{1}{p^2}\,\mathfrak{R}\{f;\, p\} \tag{27}$$

and

$$\mathfrak{R}\{f\overset{\circ}{*}g;\, p\}=\mathfrak{R}\{f;\, p\}\,\mathfrak{R}\{g;\, p\} \tag{28}$$

hold.

Proof. Since $f,\ g\in\mathscr{C}_T$, then $Tf,\ Tg\in\mathscr{C}_{\exp}$ and hence $Tf*Tg\in\mathscr{C}_{\exp}$. Therefore $f\overset{\circ}{*}g\in\mathscr{C}_T$. Then

$$\mathfrak{R}\{f\overset{\circ}{*}g;\, p\}=\mathfrak{L}\circ T[T^{-1}((Tf)*(Tg))],$$

$$=\mathfrak{L}((Tf)*(Tg))=(\mathfrak{L}\circ Tf)\,.\,(\mathfrak{L}\circ Tg)=\mathfrak{R}\{f;\, p\}\,.\,\mathfrak{R}\{g;\, p\}.$$

Thus (28) is proved. Then (27) can be obtained as a special case of (28). Indeed, since by (21) $Lf=u_2\overset{\circ}{*}f\in\mathscr{C}_T$, due to $u_2=T^{-1}\{t\}\in\mathscr{C}_T$, then, using the properties of the Laplace transform, we get

$$\mathfrak{R}\{Lf;\, p\}=\mathfrak{L}\{TLf;\, p\}=\mathfrak{L}\{t^2Tf;\, p\}=\frac{1}{p^2}\,\mathfrak{L}\{Tf;\, p\}=\frac{1}{p^2}\mathfrak{R}\{f;\, p\}.\ \square$$

Corollary. Let $\ker D\subset\mathscr{C}_T$ and let f be such a function from $\mathscr{C}^2([0,\infty))$ that $Df\in\mathscr{C}_T$. Then $f\in\mathscr{C}_T$, and the identity

$$\mathfrak{R}\{Df;\, p\}=p^2\mathfrak{R}\{f;\, p\}-f(0)\mathfrak{R}\{u_1;\, p\}-f'(0)\mathfrak{R}\{u_2;\, p\} \tag{29}$$

holds.

The proof follows from the identity $LDf=f-f(0)u_1-f'(0)u_2$ and from (27).

In order to be able to verify effectively the hypothesis of the above theorem, the following remark can be useful. If

$$\sup_{0\leq t<\infty}\left|\int\limits_0^t q(\tau)d\tau\right|<\infty, \tag{30}$$

then the similarity T maps $\mathscr{C}_{\exp}$ into itself, and hence $\mathscr{C}_T=\mathscr{C}_{\exp}$. In order to show this, we may use an estimate for the kernel $K(t,\ \tau)$ of T, given in Marchenko [74], p. 26.

3.2.2. Convolutions of the initial value right inverses of the Bessel differential operators. Next we shall consider a classical example of a linear differential operator of the second order with a singularity. We shall find convolutions of the initial value right inverse of the Bessel differential operator

$$B_\nu=\frac{d^2}{dt^2}+\frac{1}{t}\,\frac{d}{dt}-\frac{\nu^2}{t^2}\ =t^{-\nu-1}\,\frac{d}{dt}\,t^{2\nu+1}\frac{d}{dt}\,t^{-\nu} \tag{31}$$

with $\nu\geq 0$ for the singular point $t=0$.

The basic functional space in which we shall define the initial value right inverse L_ν of B_ν consists of the functions $f(t)$, defined for $0<t<\infty$ and having a representation of the form

(32) $$f(t)=t^p\tilde{f}(t)$$

with a function $\tilde{f}(t)$, defined and continuous for $0\leq t<\infty$ and with an arbitrary $p>\nu-2$. We denote this space by $\mathscr{C}_{\nu-2}$. If $\tilde{f}(t)\in\mathscr{C}^k$, then the space of such functions (32) is denoted by $\mathscr{C}^k_{\nu-2}$.

Definition 1. The *initial value right inverse* of B_ν is said to be the linear operator $L_\nu:\mathscr{C}_{\nu-2}\to\mathscr{C}^2_\nu$ such that for $f\in\mathscr{C}_{\nu-2}$, the function $L_\nu f(t)$ is the solution of the initial value problem

(33) $$B_\nu L_\nu f(t)=f(t),\quad \lim_{t\to+0} t^{-\nu}L_\nu f(t)=0.$$

It is easy to see that L_ν has a representation of the form

(34) $$L_\nu f(t)=\frac{t^2}{4}\int_0^1\int_0^1 t_1^{-\nu/2}\,t_2^{\nu/2}\,f(t\sqrt{t_1t_2})dt_1dt_2.$$

Let us find the defining projector F_ν of L_ν. As we have seen in 1.3.4, the restriction of each defining projector of L_ν onto the subdomain $\mathscr{C}^2_\nu$ of the domain of B_ν is

(35) $$Ff(t)=(I-L_\nu B_\nu)f(t)=t^\nu\lim_{\tau\to+0}\tau^{-\nu}f(\tau)=0,$$

i. e. L_ν and B_ν are mutually inverse on $\mathscr{C}^2_\nu$. It is better to consider F_ν on the space $\tilde{\mathscr{C}}^2_\nu$ of the functions of the form

(36) $$f(t)=f_0t^\nu+\tilde{f}(t)$$

with a constant f_0 and with a function $\tilde{f}(t)\in\mathscr{C}^2_\nu$. If $f\in\mathscr{C}^2_\nu$, then

(37) $$F_\nu f(t)=f_0t^\nu\quad\text{with}\quad f_0=\lim_{\tau\to+0}\tau^{-\nu}f(\tau).$$

Theorem 3.2.5 (Dimovski [43]) *The linear transform* $T_\nu:\mathscr{C}_{\nu-2}\to\mathscr{C}^{[\nu+1/2]}_{2\nu-1}\subset\mathscr{C}_{-1}$, *defined by*

(38) $$T_\nu f(t)=\frac{t^{\nu+1}}{\Gamma(\nu+1/2)}\int_0^1(1-\tau)^{\nu-1/2}\tau^{-\nu/2}f(t\sqrt{\tau})d\tau$$

is a similarity of L_ν *to* l^2, *i. e.*

(39) $$T_\nu L_\nu=l^2T_\nu$$

where l *is the Volterra integration operator* $lf(t)=\int_0^t f(\tau)d\tau$.

Proof. Let us note that here the operator l^2 is considered as defined not on its "natural" domain $\mathscr{C}_{-1}([0,\infty))$ but on the subspace $\mathscr{C}^{[\nu+1/2]}_{2\nu-1}$ of

$\mathscr{C}_{-1}([0,\infty))$. We verify the desired identity $T_\nu L_\nu f(t)=l^2 T_\nu f(t)$ at first for powers of t, i. e. for functions of the form $f(t)=t^p$ with $p>\nu-2$. Then let $f(t)=t^p\tilde{f}(t)$ with $p>\nu-2$ and with $\tilde{f}\in\mathscr{C}([0,\infty))$. Let $\tilde{f}_n(t)=\sum_{k=0}^{n} a_{nk}t^k$, $n=1, 2,\ldots$, be an almost uniformly convergent sequence to $\tilde{f}(t)$ in $\mathscr{C}([0,\infty])$. If, $f_n(t)=t^p\tilde{f}_n(t)$, then $T_\nu L_\nu f_n(t)=l^2 T_\nu f_n(t)$ for $n=1, 2,\ldots$. For $n\to\infty$, we get relation (39). □

In non-explicit form transformation (38) with $|\nu|<1/2$ can be seen in the so-called "Sonine integrals", which transform Bessel functions into sine and cosine. That is why, following Delsarte [28], p. 439, we name (38) the *Sonine transform.*

Lemma 4. *The Sonine transform T_ν is invertible, and its inverse transform T_ν^{-1} has a representation of the form*

$$(40)\qquad T_\nu^{-1}f(t)=\frac{t^\nu}{2^{\nu+1/2}}\left(\frac{1}{t}\frac{d}{dt}\right)^{\nu+1/2}\{f(t)\}$$

when $\nu+1/2$ is an integer, i. e. when ν is half of an odd integer, and in the form

$$(41)\qquad T_\nu^{-1}f(t)=\frac{t^\nu}{2^{[\nu+1/2]+1}\Gamma(1-\{\nu+1/2\})}\left(\frac{1}{t}\frac{d}{dt}\right)^{[\nu+1/2]+1}\int_0^1(1-\tau)^{-\{\nu+1/2\}}f(t\sqrt{\tau})d\tau$$

when $\nu+1/2$ is not an integer, where by $[\nu+1/2]$ and $\{\nu+1/2\}$ are denoted the integer and the fractional parts of $\nu+1/2$ correspondingly. In the first case T_ν^{-1} is defined on $\mathscr{C}_{2\nu-1}^{\nu+1/2}$, and in the second in $\mathscr{C}_{2\nu-1}^{[\nu+1/2]+1}$.

Proof. The invertibility of T_ν is a simple consequence from the Titchmarsh theorem on convolution (see 1.1). Indeed the equation $T_\nu\{g(t)\}=0$ can be written in the form $\{t^{\nu-1/2}\}*\{t^{-\nu/2}g(\sqrt{t})\}=0$, where $*$ denotes the Duhamel convolution. Therefore, $g(\sqrt{t})=0$, even when we consider $*$ in a finite interval $[0, A]$, since $t^{\nu-1/2}$ does not vanish in any right neighbourhood of $t=0$. The function $T_\nu^{-1}f(t)=g(t)$ we search as the solution of the first kind integral equation $T_\nu y=f(t)$, which can easily be written in the form $l^{\nu+1/2}\{t^{-\nu/2}y(\sqrt{t})\}=f(\sqrt{t})$. The formal solution of this equation is

$$y(\sqrt{t})=t^{\nu/2}\left(\frac{d}{dt}\right)^{\nu+1/2}\{f(\sqrt{t})\},$$

where $(d/dt)^{\nu+1/2}$ is the operator of fractional differentiation of order $\nu+1/2$ when $\nu+1/2$ is not an integer, and of ordinary differentiation of order $\nu+1/2$, when it is an integer. Formula (41) is obtained using the properties of the operator $(d/dt)^{\nu+1/2}$. □

Having at our disposal the similarity T_ν of L_ν to l^2 and using the results of 1.3.3, we can at once write down a convolution of L_ν in $\mathscr{C}_{\nu-2}$. This is the operation

$$(42)\qquad (f\overset{\nu}{*}g)(t)=T_\nu^{-1}[(T_\nu f)*(T_\nu g)].$$

In the following theorem an explicit representation of (42) is given.

Theorem 3.2.5. *Convolution* (42) *admits a representation of the form*

$$f\overset{\nu}{*}g=\Lambda_\nu(f\times g) \tag{43}$$

with

$$(f\times g)(t)=t^2\int_0^1\int_0^1\left[\frac{t_1(1-t_1)}{t_2(1-t_2)}\right]^{\nu/2}f(t\sqrt{t_1t_2})g(t\sqrt{(1-t_1)(1-t_2)})dt_1dt_2 \tag{44}$$

and with the operator $\Lambda_\nu:\mathscr{C}_{2\nu-2}\to\mathscr{C}_{3\nu-2}$, *defined by*

$$\Lambda_\nu h(t)=\begin{cases}\dfrac{\sqrt{\pi}\,t_\nu}{2^{\nu+1}\Gamma(\nu)}\displaystyle\int_0^1(1-\tau)^{\nu-1}\tau^{-\nu}h(t\sqrt{\tau})d\tau & \text{for } \nu>0,\\[2ex] \dfrac{\sqrt{\pi}}{2}\,h(t) & \text{for } \nu=0.\end{cases} \tag{45}$$

Proof. We verify identity (43) at first for powers of t, i. e. for functions of the form $f(t)=t^p$ and $g(t)=t^q$ with $p,q>\nu-2$. If now $f,g\in\mathscr{C}_{\nu-2}$ are arbitrary functions of the form $f(t)=t^p\tilde{f}(t)$ and $g(t)=t^q\tilde{g}(t)$ with $\tilde{f}(t)$ $\tilde{g}(t)\in\mathscr{C}([0,\infty))$, then we take polynomial sequences $\{\tilde{f}_n(t)\}_{n=1}^{\infty}$ and $\{g_n(t)\}_{n=1}^{\infty}$ converging almost uniformly to $\tilde{f}(t)$ and $\tilde{g}(t)$. correspondingly. If $n\to\infty$ in the already proved identity $f_n\overset{\nu}{*}g_n=\Lambda_\nu(f_n\times g_n)$ with $f_n(t)=t^p\tilde{f}_n(t)$ and $g_n(t)=t^q\tilde{g}_n(t)$, then (43) follows using the continuity of Λ_ν and $\times$.

Representation (43) is an interesting one, since it contains fewer integrations than (42).

Now, having at our disposal the similarity T_ν of L_ν to l^2 along with the explicit representation (43) of the convolution (42) of L_ν in $\mathscr{C}_{\nu-2}$, without any difficulties the elements of the corresponding operational calculus can be developed. Let S_ν stand for the inverse element of L_ν in the multiplier quotients ring $\mathfrak{M}_\nu$ of convolution (42), i. e. $S_\nu=1/L_\nu$. Let us find the connection between B_ν and S_ν. From expression (35) of the defining projector F_ν of L_ν in $\tilde{\mathscr{C}}_\nu^2$, we get

$$B_\nu f=S_\nu f-f_0S_\nu\{t^\nu\} \tag{46}$$

with $f_0=\lim\limits_{\tau\to+0}\tau^{-\nu}f(\tau)$. For an effective use of the corresponding operational calculus, we need explicit expressions for the partial fractions of the form $S_\nu\{t^\nu\}/(S_\nu-\lambda)^n$. It is not difficult to show that

$$\frac{S_\nu\{t^\nu\}}{S_\nu-\lambda}=\{J_\nu(\sqrt{-\lambda}\,t\}* \tag{47}$$

and

$$\frac{S_\nu\{t^\nu\}}{(S_\nu-\lambda)^n}=\left\{\frac{1}{(n-1)!}\frac{\partial^{n-1}}{\partial\lambda^{n-1}}J_\nu(\sqrt{-\lambda}\,t)\right\}*\quad\text{for}\quad n\geq1. \tag{48}$$

Example. In axially symmetric problems of mathematical physics an important role is played by the Bessel differential operator

$$B_0=\frac{d^2}{dt^2}+\frac{1}{t}\frac{d}{dt} \tag{49}$$

with $\nu=0$. We are interested in the initial value right inverse operator L_0 of B_0, given by

$$L_0f(t)=\frac{t^2}{4}\int_0^1\int_0^1 f(t\sqrt{t_1t_2})\,dt_1dt_2. \tag{50}$$

Then the corresponding Sonine transformation, serving as a similarity of L_0 to l^2, has the form

$$T_0f(t)=\frac{2}{\sqrt{\pi}}\int_0^t \frac{\tau f(\tau)d\tau}{\sqrt{t^2-\tau^2}}. \tag{51}$$

Since in this case Λ_0 from (45) is a multiplication with the constant $\sqrt{\pi}/2$, i. e. $\Lambda_0h(t)=(\sqrt{\pi}/2)h(t)$, then convolution (43) of L_0 in $\mathscr{C}_{-2}$ takes the form

$$(f\overset{\circ}{*}g)(t)=\frac{\sqrt{\pi}\,t^2}{2}\int_0^1\int_0^1 f(t\sqrt{t_1t_2})\,g(t\sqrt{(1-t_1)(1-t_2)})\,dt_1dt_2. \tag{52}$$

Here the defining projector F_0 of L_0 is more appropriate to be considered in the space $\tilde{\mathscr{C}}_0^2$ of the functions of the form

$$f(t)=f_0\ln t+\tilde{f}(t)$$

with constants f_0 and with $\tilde{f}(t)\in\mathscr{C}_0^2$. Then

$$F_0f(t)=f_0\ln t \quad\text{with}\quad f_0=\lim_{\tau\to+0}\frac{f(\tau)}{\ln\tau}. \tag{53}$$

Two types of partial fractions with S_0 are of interest:

$$\frac{1}{S_0-\lambda}=\left\{\frac{1}{2\sqrt{\pi}}J_0(\sqrt{-\lambda}\,t)\right\}* \tag{54}$$

and

$$\frac{S_0\{\ln t\}}{S_0-\lambda}=\left\{\frac{\pi}{2}Y_0(\sqrt{-\lambda}\,t)+\left(\ln\frac{2}{\sqrt{-\lambda}}-\gamma\right)J_0(\sqrt{-\lambda}\,t)\right\}*, \tag{55}$$

where $\gamma=0.577215\ldots$ is the Euler constant.

The corresponding partial fractions $1/(S_0-\lambda)^n$ and $S_0\{\ln t\}/(S_0-\lambda)^n$ with integer $n>1$, can be obtained from (54) and (55) by a formal differentiation with respect to λ.

In the same way as in the previous subsection, an integral transformation connected with the initial value right inverse operator of the general non-singular linear differential operator of the second order had been introduced, as for the Bessel differential operator there exists analogous integral transformation. This is the Meijer-Obrechkoff integral transformation (Obrechkoff [83]).

Definition. The *Meijer-Obrechkoff transformation* is said to be the composition of the Laplace transform and the Sonine operator (38), i. e. the integral transformation

$$\mathfrak{K}_\nu\{f(t);\, p\}=\int\limits_0^\infty e^{-pt}(T_\nu f)(t)\,dt. \tag{56}$$

We consider the Meijer-Obrechkoff transformation in the space $\mathscr{C}_{\nu-2}^{\exp}$ of the functions of $\mathscr{C}_{\nu-2}$, which are $O(e^{\alpha t})$ for $t\to+\infty$ with real α.

Taking into account some elementary properties of the Laplace transform, and identities (39) and (42), it is easy to show (see Dimovski [33]) the following operational properties of the Meijer-Obrechkoff integral transformation:

$$\mathfrak{K}_\nu\{L_\nu f(t);\, p\}=\frac{1}{p^2}\,\mathfrak{K}_\nu\{f(t);\, p\} \tag{57}$$

and

$$\mathfrak{K}_\nu\{(f\overset{\nu}{*}g)(t);\, p\}=\mathfrak{K}_\nu\{f(t);\, p\}\,\mathfrak{K}_\nu\{g(t);\, p\}. \tag{58}$$

Usually, *Meijer transformation* is said to be the integral transformation

$$\tilde{\mathfrak{K}}_\nu\{f(t);\, p\}=\int\limits_0^\infty tK_\nu(pt)f(t)\,dt.$$

In 1954 Obrechkoff [83], p. 85, studied its modification

$$\tilde{\tilde{\mathfrak{K}}}_\nu\{f(t);\, p\}=p^{-\nu}\int\limits_0^\infty tK_\nu(pt)f(t)dt.$$

We shall show that this modification up to a numerical multiplier coincides with (56).

Theorem 3.2.6. *The representation*

$$\mathfrak{K}_\nu\{f(t);\, p\}=\frac{2^{\nu+1}}{\sqrt{\pi}}\,p^{-\nu}\int\limits_0^\infty tK_\nu(pt)f(t)\,dt, \tag{59}$$

where $K_\nu(x)$ is the MacDonald function, holds.

Proof. First, we shall show the validity of (59) for real $p>0$, and then its validity for complex p shall follow from the principle of analytic continuation. Writing the Sonine transformation T_ν in the form

$$T_\nu f(t)=\frac{2}{\Gamma(\nu+1/2)}\int\limits_0^t(t^2-\tau^2)^{\nu-1/2}\tau^{1-\nu}f(\tau)\,d\tau \tag{60}$$

and changing the order of the integration in (56), we get

$$\mathfrak{K}_\nu\{f(t);\,p\}=\frac{2}{\Gamma(\nu+1/2)}\int\limits_0^\infty \tau^{1-\nu}f(\tau)\,dt\int\limits_\tau^\infty (t^2-\tau^2)^{\nu-1/2}\,e^{-pt}\,dt$$

$$=\frac{2}{\Gamma(\nu+1/2)}\int\limits_0^\infty \tau^{1+\nu}f(\tau)\,d\tau\int\limits_0^\infty (\sigma^2-1)^{\nu-1/2}e^{-p\sigma\tau}d\sigma.$$

It remains to use the well-known integral representation

$$\Gamma(\nu+1/2)K_\nu(z)=\sqrt{\pi}\left(\frac{z}{2}\right)^\nu\int\limits_1^\infty e^{-zt}(t^2-1)^{\nu-1/2}\,dt,\ \operatorname{Re}\nu>-\frac{1}{2},\ \operatorname{Re}z>0$$

of the function $K_\nu(z)$ (see Erdelyi et al. [56], p. 82) in order to get (59).

The Meijer-Obrechkoff transformation can be used for a functional development of the operational calculus for the initial value right inverse operator L_ν of B_ν.

A convolution of the integral transformation

$$\hat{\mathfrak{K}}_\nu\{f(t);\,p\}=2\int\limits_0^\infty (pt)^{\nu/2}K_\nu(2\sqrt{pt})f(t)\,dt,$$

related to the Meijer-Obrechkoff transformation, had been found by Krätzel [71] in 1965.

3.3. CONVOLUTIONS OF BOUNDARY VALUE RIGHT INVERSES OF LINEAR SECOND-ORDER DIFFERENTIAL OPERATORS

In this section convolutions for some general classes of right inverse operators of an arbitrary non-singular linear differential operator are found. The boundary value problems connected with these right inverse operators include as a special case the classical Sturm-Liouville eigenvalue problem and boundary value problems with non-local boundary condition. That is why specializations of these convolutions can be used as convolutions for the finite Sturm-Liouville integral transformations. As an example of a singular second-order linear differential operator the Bessel differential operator is considered. Thus convolutions for the classical finite Bessel transformations are found.

3.3.1. Convolutions of right inverses of non-singular second-order linear differential operator, determined by a Sturm-Liouville and a general boundary value conditions. As in the first part of the previous section we consider again an arbitrary linear differential operator of the second order

(1)
$$D=a(x)\frac{d^2}{dx^2}+b(x)\frac{d}{dx}+c(x)$$

in a segment $\Delta=[a,\,b]$ under the assumption that

$$a(x)\in\mathscr{C}^2(\Delta),\ b(x)\in\mathscr{C}^1(\Delta)\ \text{and}\ c(x)\in\mathscr{C}(\Delta).$$

Let $\Phi: \mathscr{C}^1(\Delta) \to \mathbf{C}$ be an arbitrary non-zero linear functional in $\mathscr{C}^1(\Delta)$ continuous in the norm of $\mathscr{C}^1(\Delta)$: $\|f\|_{C^1} = \max_{x \in \Delta} \{|f(x)| + |f'(x)|\}$. The right inverse operator $L: \mathscr{C}(\Delta) \to \mathscr{C}(\Delta)$ of D in $\mathscr{C}(\Delta)$ we are to consider, is defined in such a way that $y = Lf$ is the solution of the elementary boundary value problem

(2) $$Dy = f, \quad \alpha y(a) + \beta y'(a) = 0, \quad \Phi(y) = 0,$$

where α and β are numbers which are not both zero. In order problem (2) to have a solution for each $f \in \mathscr{C}([a, b])$, it is necessary and sufficient $\Phi(\alpha y_2 - \beta y_1) \neq 0$, where y_1, y_2 is the fundamental system of D for the point $x = a$, i. e. $Dy_i = 0$, $i = 1, 2$; $y_1(a) = 1$, $y_1'(a) = 0$, $y_2(a) = 0$, $y_2'(a) = 1$. Then the right inverse L of D we are interested in can be represented in the form

(3) $$Lf = L_0 f - y\, \Phi(L_0 f),$$

where L_0 is the initial value right inverse of D for the point $x = a$, i. e. L_0 is defined by $DL_0 f = f$ and $(L_0 f)(a) = (L_0 f)'(a) = 0$. Here y stands for $(\alpha y_2 - \beta y_1)/\Phi\{\alpha y_2 - \beta y_1\}$.

Here we shall show that L has a non-trivial convolution in $\mathscr{C}([a, b])$ and we shall find an explicit representation of such a convolution. In this we follow Bozhinov and Dimovski [18, 19]. Using the similarity method, we reduce this problem to an analogous problem, but for the square of the differentiation, already solved in sect. 3.1. As in the previous section, it can easily be seen that instead of the general operator (1) we can consider, without an essential loss of generality, the operator

(4) $$D = \frac{d^2}{dt^2} - q(t), \quad q \in \mathscr{C}([0, 1]),$$

in the standard segment [0, 1]. Indeed, after the analogous change of the independent variable by $t = \varphi(x) = \int_a^x \frac{d\xi}{\sqrt{a(\xi)}} \Big/ \int_a^b \frac{d\xi}{\sqrt{a(\xi)}}$, the segment $[a, b]$ goes into [0, 1]. Further, by a well-known substitution for the function, the coefficient of d/dx in the operator D can vanish. Thus we get an operator of the form (4), similar to the given operator (1) by a similarity of the change of the variables type.

Henceforth we consider operator (4) and boundary value problem

(5) $$DLf(t) = f(t), \quad \alpha(Lf)(0) + \beta(Lf)'(0) = 0, \quad \Phi(Lf) = 0,$$

for the right inverse L of D in $\mathscr{C}([0, 1])$. Here Φ is again a non-zero linear functional in $\mathscr{C}^1([0, 1])$, continuous in the norm

(6) $$\|f\|_{C^1} = \max_{0 \le x \le 1} \{|f(x)| + |f'(x)|\}.$$

We suppose that if $u_1(t)$, $u_2(t)$ form the fundamental system for D in the point $t = 0$, then the condition $\omega = \Phi(\beta u_1 - \alpha u_2) \neq 0$ for solvability of (5) is fulfilled. If we denote $u(t) = (\beta u_1 - \alpha u_2)/\omega$, then L has the representation

(7) $$Lf(t) = L_0 f(t) - u(t)\, \Phi(L_0 f),$$

where L_0 is the initial value right inverse of D for the point $t=0$. We shall show that L is similar to a right inverse operator of d^2/dt^2 of the type considered in sect. 3.1. To this end, we should distinguish the two cases: (a) $\beta \neq 0$; (b) $\beta = 0$.

(a) $\beta \neq 0$.

Lemma 1. *If $\beta \neq 0$, then there exists a function $K(t,\tau)$, $K \in \mathscr{C}^1(G)$, where $G=\{(t,\tau): 0 \leq t \leq 1, 0 \leq \tau \leq t\}$, such that the second kind Volterra transformation*

$$(8) \qquad Tf(t)=f(t)+\int_0^t K(t,\tau)f(\tau)d\tau$$

is a similarity of L, given by (7) to the right inverse operator

$$(9) \qquad \tilde{L}f(t)=l^2 f(t)-\frac{\beta}{\omega}\tilde{\Phi}(l^2 f)$$

of d^2/dt^2 with $\tilde{\Phi}=\Phi \circ T^{-1}$, i.e. $TL=\tilde{L}T$.

Proof. As in 3.2.1, at first we consider the case of smooth $q(t)$, i. e. $q \in \mathscr{C}^1([0,1])$. Then let $K(t,\tau)$ be the solution of the partial differential equation

$$(10) \qquad K_{tt}-K_{\tau\tau}+q(\tau)K=0$$

in the domain $G=\{(t,\tau): 0 \leq t \leq 1,\ 0 \leq \tau \leq t\}$ and satisfying the boundary value conditions

$$(11) \qquad K(t,t)=-h-\frac{1}{2}\int_0^t q(\tau)d\tau$$

and

$$(12) \qquad K_\tau(t,0)-hK(t,0)=0$$

with $h=-\alpha/\beta$. In the same way as in 3.2.1 the characteristic substitution $u=(t+\tau)/2$, $v=(t-\tau)/2$, reduces the problem (10), (11), (12) to the problem

$$k_{uv}=-q(u-v)k,$$

$$(13) \qquad k(u,0)=-h-\frac{1}{2}\int_0^u q(\xi)d\xi,$$

$$k_u(u,u)-k_v(u,u)-hk(u,u)=0$$

for the function $k(u,v)=K(u+v,\ u-v)$.

By means of successive integration on u and v of the partial differential equation (13), and using the boundary conditions, we get the following second kind Volterra integral equation

$$(14) \qquad k(u,v)=-he^{-2hu}-e^{2hv}\int_0^v e^{2h\xi}q(\xi)d\xi-\frac{1}{2}\int_v^u q(\xi)d\xi$$

$$-e^{-2hv}\int_0^v e^{2h\xi}\,d\xi\int_0^\xi q(\xi-\eta)k(\xi,\eta)\,d\eta-\int_v^u d\xi\int_0^v q(\xi-\eta)k(\xi,\eta)d\eta.$$

Since this integral equation always has a solution $k(u, v)$ in the whole domain, then in reverse order it can be seen that the function $K(t, \tau) = k((t+\tau)/2, (t-\tau)/2)$ is a solution of boundary value problem (10), (11), (12). In the same way as in 3.2.1, it is seen that for each $f \in \mathscr{C}^2([0, 1])$ the identity

$$TD\,f(t)=\frac{d^2}{dt^2}T\,f(t)-(f'(0)-hf(0))K(t,0) \tag{15}$$

holds. Substituting $L\,f(t)$ instead of $f(t)$, and using the defining condition $(L\,f)'(0)-h(L\,f)(0)=0$ for L, we get

$$\frac{d^2}{dt^2}TL\,f(t)=T\,f(t). \tag{16}$$

From (16) the similarity relation $TL=\tilde{L}T$ follows immediately. Thus the assertion is proved in the case $\beta \neq 0$. In the same way as in the proof of Lemma 3 in sect. 3.2 it is seen that in this case the integral equation (15) has a smooth solution $k(u, v)$ in the corresponding domain and that transformation (8) with the kernel $K(t, \tau)=k((t+\tau)/2, (t-\tau)/2)$ is the desired similarity. This can be shown in a standard manner by approximation of $q(t)$ by functions $q_n(t) \in \mathscr{C}^1([0, 1])$ in the topology of $\mathscr{C}([0, 1])$. □

Theorem 3.3.1. *The operation*

$$(f*g)(t)=T^{-1}\circ l\left(\int_0^t (Tf)(t-\tau)(Tg)(\tau)\,d\tau\right) \tag{17}$$

$$-\frac{\beta}{2\omega}\Phi_x\left\{T_t^{-1}\circ T_x^{-1}\circ l_x\left(\int_t^x (Tf)(x+t-\tau)(Tg)(\tau)d\tau\right.\right.$$

$$\left.\left.+\int_{-t}^x (Tf)(|x-t-\tau|)(Tg)(|\tau|)\,d\tau\right)\right\}$$

is a convolution of L *in* $\mathscr{C}([0, 1])$.

Proof. By Theorem 1.3.6, the operation $f*g=T^{-1}[(Tf)\,\tilde{*}\,(Tg)]$, where

$$(f\,\tilde{*}\,g)(t)=\int_0^t d\tau\int_0^\tau f(\tau-\sigma)g(\sigma)\,d\sigma-\frac{\beta}{2\omega}\Phi_x\left\{T_x^{-1}\int_0^x d\xi\left(\int_t^\xi f(\xi+t-\tau)g(\tau)\,d\tau\right.\right.$$

$$\left.\left.+\int_{-t}^x f(|x-t-\tau|)g(|\tau|)\,d\tau\right)\right\}$$ is a continuous convolution of (9) in $\mathscr{C}([0, 1])$

Obviously, it coincides with operation (17). □

Corollary 1. The operator L has the representation

$$L\,f=\left\{-\frac{\omega}{\beta}\,u(t)\right\}*f \tag{18}$$

as a convolutional operator.

Indeed, since $Tu(t)=-\beta/\omega=\text{const}$, and $\tilde{L}f=\{1\}\tilde{*}f$, then $Lf=(T^{-1}\{1\})*f=\left\{-\frac{\omega}{\beta}u\right\}*f$.

(b) $\beta=0$.

Then the right inverse L of D is defined as the solution of the boundary value problem

(19) $$DLf(t)=f(t),\ (Lf)(0)=0,\ \Phi(Lf)=0$$

and it has the representation

(20) $$Lf(t)=L_0f(t)-u_2(t)\,\Phi(L_0f),$$

where without a loss of generality it is assumed that $\Phi(u_2)=1$. Let us remind that $u_2(t)$ is the solution of $Du=0$ with $u(0)=0$, $u'(0)=1$.

Lemma 2. *If $q(t)\in\mathscr{C}([0,1])$, then there exists a function $K(t,\tau)$ from $\mathscr{C}^1(G)$, where $G=\{(t,\tau):0\leqq t\leqq 1, 0\leqq\tau\leqq t\}$, such that the second kind Volterra transformation*

(21) $$Tf(t)=f(t)+\int_0^t K(t,\tau)f(\tau)d\tau$$

is a similarity of the operator L to the right inverse operator

(22) $$\tilde{L}f(t)=l^2f(t)-\tilde{\Phi}(l^2f)t,\quad \tilde{\Phi}=\Phi\circ T^{-1}$$

of d^2/dt^2.

Proof. As in Lemma 1, first we consider the case $q(t)\in\mathscr{C}^1([0,1])$ and define $K(t,\tau)$ as the solution of the boundary value problem

(23) $$K_{tt}-K_{\tau\tau}+q(\tau)K=0$$
$$K(t,t)=-\frac{1}{2}\int_0^t q(\xi)d\xi,\quad K_\tau(t,0)=0$$

in the domain G. In the same way problem (23) is reducible to the second kind Volterra integral equation

(24) $$k(u,v)=-\frac{1}{2}\int_0^u q(\xi)d\xi-\frac{1}{2}\int_0^v q(\xi)d\xi$$
$$-\int_0^v d\xi\int_0^\xi q(\xi-\eta)k(\xi,\eta)d\eta-\int_v^u d\xi\int_0^\xi q(\xi-\eta)k(\xi,\eta)d\eta$$

for the function $k(u,v)=K(u+v,u-v)$. But (24) has a smooth solution for $q(t)\in\mathscr{C}([0,1])t_{00}$. Then the function $K(t,\tau)=k((t+\tau)/2,(t-\tau)/2)$ is the kernel of a similarity (21) of the desired kind. □

Theorem 3.3.2. *The operation*

$$(25)\qquad (f*g)(t)=-\frac{1}{2}\,\Phi_x\Big\{T_t^{-1}\circ T_x^{-1}\circ l_x\Big(\int_t^x (Tf)(x+t-\tau)\,(Tg)(\tau)d\tau$$

$$-\int_{-t}^{x}(Tf)(|x-t-\tau|)(Tg)(|\tau|)\,\mathrm{sgn}\,((x-t-\tau)\tau)d\tau\Big)\Big\}$$

is a continuous convolution of the operator (20) *in* $\mathscr{C}([0,1])$.

Proof. By Theorem 1.3.6, the operation $f*g=T^{-1}[(Tf)\,\tilde{*}\,(Tg)]$, with

$$(f\,\tilde{*}\,g)(t)=-\frac{1}{2}\Phi_x\circ T_x^{-1}\circ l_x\Big\{\int_t^x f(x+t-\tau)g(\tau)d\tau-\int_{-t}^{x}f(|x-t-\tau|)\,g(|\tau|)\times$$

$$\times\mathrm{sgn}((x-t-\tau)\tau)d\tau\Big\}$$ is a convolution of L in $\mathscr{C}([0,1])$. This is exactly operation (25). □

Corollary. The operator L has the convolutional representation

$$(26)\qquad Lf=\{u_2(t)\}*f.$$

The convolutions (17) and (25) give a complete solution of the problem stated in the beginning of this section. They can be used for developing operational calculi for corresponding boundary value problems. If $S=1/L$ is the inverse element of L in the multiplier quotients ring of a convolution of the kind considered, then the first problem in the operational calculus for L is the establishing of the relation between Df and Sf for arbitrary $f\in\mathscr{C}^2([0,1])$.

Theorem 3.3.3. *Let* $f\in\mathscr{C}^2([0,1])$. *Then in the multiplier quotients rings of the convolutions* (17) *and* (25) *the relation*

$$(27)\qquad Df=Sf-\frac{\beta}{\omega}\Phi(f)-[\alpha f(0)+\beta f'(0)]S\{v(t)\},$$

with $v(t)=\frac{1}{2}[\Phi(u_1)u_2(t)-\Phi(u_2)u_1(t)]$ *when* $\beta\neq 0$, *and the relation*

$$(28)\qquad Df=Sf-\Phi(f)-f(0)S\{v(t)\},$$

with $v(t)=u_1(t)-\Phi(u_1)u_2(t)$ *when* $\beta=0$, *holds.*

Proof. It is easy to see that in both cases the defining projector F of the right inverse operator L of D has the form

$$(29)\qquad Ff=\Phi(f)u+[\alpha f(0)+\beta f'(0)]v,$$

where in the first case $u=(\beta u_1-\alpha u_2)/\omega$, $v=(\Phi(u_1)u_2-\Phi(u_2)u_1)/\omega$, and $u=u_2$, $v=u_1-u_2\Phi(u_1)$ in the second case. Taking into account representations (18) and (26), the desired relations (27) and (28) are obtained at once from (29). □

Example. Let $\Phi(f)=\gamma f(1)+\delta f'(1)$. Then the right inverse operator L of D is closely connected with the classical *Sturm-Liouville problem.* In the next subsection we shall consider in more detail some applications of the convolutions obtained for this problem.

Now we shall show briefly the possibility of extending the convolutions (17) and (25) as continuous convolutions of the operators (9) and (20) in the whole space $\mathscr{L}([0,1])$.

Lemma 3. *Let Φ be a bounded operator in $\mathscr{C}([0,1])$. Then the operation*

$$(f*g)(t)=\Phi_x\left\{\int_x^t f(t+x-\tau)g(\tau)d\tau\right\} \tag{30}$$

admits an estimate of the form

$$\|f*g\|_1\leq A\|f\|_1\cdot\|g\|_1 \tag{31}$$

for $f,g\in\mathscr{C}([0,1])$, with $A=\|\Phi\|_{C([0,1])}$, where $\|f\|_1=\int_0^1|f(\tau)|\,d\tau$ denotes the integral norm.

Proof. If $\Phi(f)=\int_0^1 f(\tau)d\mu(\tau)$ is the Riesz representation of Φ with a complex measure $\mu(\tau)=\mu_1(\tau)+i\mu_2(\tau)$, then let us denote $|\Phi|(f)=\int_0^1 f(\tau)\,d|\mu|(\tau)$ with $|\mu|(\tau)=|\mu_1|(\tau)+i|\mu_2|(\tau)$, where $|\mu_1|(\tau)$ and $|\mu_2|(\tau)$ stand for the variation of μ_1 and μ_2 on $[0,\tau]$. Then we get

$$\|f*g\|_1=\int_0^1|(f*g)|(t)dt\leq|\Phi|_x\left\{\int_0^1\operatorname{sgn}(t-x)\,dt\int_t^x|f(t+x-\tau)|.|g(\tau)|\,d\tau\right\}$$

$$=|\Phi|_x\left\{\int_0^x dt\int_t^x|f(x+t-\tau)|.|g(\tau)|d\tau+\int_x^1 dt\int_x^t|f(x+t-\tau)|.|g(\tau)|\,d\tau\right\}$$

$$\leq|\Phi|_x\left\{\int_0^x|g(\tau)|\,d\tau\int_0^\tau|f(x+t-\tau)|\,dt+\int_x^1|g(\tau)|\,d\tau\int_\tau^1|f(x+t-\tau)|\,dt\right\}.$$

Since

$$\int_0^\tau|f(x+t-\tau)|\,dt=\int_{x-\tau}^x|f(u)|\,du\leq\int_0^1|f(u)|\,du=\|f\|_1$$

and

$$\int_\tau^1|f(x+t-\tau)|\,dt=\int_x^{1+x-\tau}|f(u)|\,du\leq\int_0^1|f(u)|\,du=\|f\|_1,$$

then

$$\|f*g\|_1\leq|\Phi|_x\left\{\left(\int_0^x|g(\tau)|\,d\tau+\int_x^1|g(\tau)|\,d\tau\right).\|f\|_1\right\}=\|f\|_1\cdot\|g\|_1\cdot|\Phi|_x(\{1\}).$$

But $|\Phi_x|(1)=\|\Phi\|_{C([0,1])}$, and hence (31) is proved. □

Theorem 3.3.4. *If* $*$ *is one of operations* (17) *or* (25), *then for* $f, g \in \mathscr{C}([0, 1])$ *the estimate*

$$\|f * g\|_1 \leq A \|f\|_1 . \|g\|_1, \tag{32}$$

with a suitable constant A, *holds. The operations* (17) *and* (25) *are extendable on the space* $\mathscr{L}([0, 1])$ *as continuous convolutions of the corresponding operator* L.

Proof. By the expressions of (17) and (25), using the boundedness of the similarities (8) and (21), considered as operators on $\mathscr{L}$ ([0, 1]), the above estimate (31) can be obtained directly from (31). The last assertion of the theorem is an immediate consequence of the estimate (32). □

3.3.2. Convolutions of the finite Sturm-Liouville integral transformations. Now let $D=(d^2/dt^2)-q(t)$ with a real-valued $q \in \mathscr{C}([0. 1])$. A *Sturm-Liouville right inverse* of D is named the operator $L: \mathscr{C}([0, 1]) \to \mathscr{C}^2([0, 1])$, defined by the solution $y=Lf$ of the elementary boundary value problem

$$\begin{aligned} &Dy=f, \\ &N_\alpha(y)=\cos\alpha\, y(0)+\sin\alpha\, y'(0)=0, \\ &N_\beta(y)=\cos\beta\, y(1)+\sin\beta\, y'(1)=0. \end{aligned} \tag{33}$$

We assume that $\lambda=0$ is not an eigenvalue of the Sturm-Liouville eigenvalue problem

$$\begin{aligned} &Dy-\lambda y=0 \\ &N_\alpha(y)=0, \quad N_\beta(y)=0, \end{aligned} \tag{34}$$

which is equivalent to the requirement L to exist on $\mathscr{C}([0, 1])$.

It is well known that the Sturm-Liouville eigenvalue problem (34) has an infinite system of eigenvalues $\lambda_1, \lambda_2, \ldots, \lambda_n, \ldots$, and to it corresponds a total system of eigenfunctions $\varphi_1, \varphi_2, \ldots, \varphi_n, \ldots$ It is total in the sense that if for a function $f \in \mathscr{L}^1([0, 1])$ we have $\int_0^1 f(t)\varphi_n(t)dt=0$, for $n=1, 2, 3, \ldots$, then $f=0$ almost everywhere in [0, 1]. In particular, if $f \in \mathscr{C}([0, 1])$, then $f=0$ identically. By means of the total system of the eigenfunctions of the Sturm-Liouville problem (34) the finite Sturm-Liouville integral transformation is introduced (see Churchill [25], p. 325).

Definition. A *finite Sturm-Liouville integral transformation* is said to be the correspondence

$$T_n\{f\}=\int_0^1 f(t)\varphi_n(t)dt, \quad n=1, 2, 3, \ldots, \tag{35}$$

which associates the Fourier coefficients sequence of each $f \in \mathscr{C}([0, 1])$ with respect to the eigenfunctions φ_n, $n=1, 2, \ldots$, of the Sturm-Liouville eigenvalue problem (34).

If φ_n is the eigenfunction of (34), corresponding to the eigenvalue λ_n, then φ_n is an eigenfunction of L, but with the eigenvalue $1/\lambda_n$, i. e.

$$L\varphi_n = \frac{1}{\lambda_n}\varphi_n. \tag{36}$$

From Theorem 1.3.3 it follows that φ_n is a divisor of zero of each convolution of L, and a divisor of zero in particular of convolutions (17) and (25). From Theorem 1.3.4 we have

$$\varphi_m * \varphi_n = 0, \quad \text{for } m \neq n, \tag{37}$$

for each convolution $*$ of L.

The basic operational property of the finite Sturm-Liouville integral transformation, connected with the eigenvalue problem (34), is the relation

$$T_n\{Df\} = \lambda_n T_n\{f\} + N'_\alpha\{\varphi_n\}N_\alpha\{f\} - N'_\beta\{\varphi_n\}N_\beta\{f\} \tag{38}$$

with

$$\begin{aligned} N'_\alpha(f) &= -\sin\alpha f(0) + \cos\alpha f'(0), \\ N'_\beta(f) &= -\sin\beta f(1) + \cos\beta f'(1) \end{aligned} \tag{39}$$

(see Churchill [25], p. 327). Substituting $Lf(t)$ instead of $f(t)$ in (38) we get

$$T_n\{Lf\} = \frac{1}{\lambda_n} T_n\{f\}, \; n = 1, 2, \ldots. \tag{40}$$

Hence the finite Sturm-Liouville integral transformation "algebraizes" the corresponding Sturm-Liouville right inverse of D.

For the sake of definiteness of the following considerations, we shall confine ourselves to the case $\cos\alpha \neq 0$ to which the convolution (17) corresponds.

Lemma 4. *For each $f \in \mathscr{C}([0, 1])$ the identities*

$$f * \varphi_n = \chi_n(f)\varphi_n, \quad n = 1, 2, 3, \ldots, \tag{41}$$

hold. Here $\{\chi_n\}$ is a system of linear functionals which are multiplicative with respect to convolution (17) *of the operator L, i. e.*

$$\chi_n(f * g) = \chi_n(f)\,\chi_n(g), \; n = 1, 2, \ldots. \tag{42}$$

The system $\{\chi_n\}$ is biorthogonal to the eigenfunctions system $\{\varphi_m\}$, i. e. $\chi_n(\varphi_m) = 0$ for $n \neq m$, and *$\chi_n(\varphi_n) \neq 0$ for each n.*

Proof. From $L\varphi_n = \frac{1}{\lambda_n}\varphi_n$ it follows $L(f * \varphi_n) = f * (L\varphi_n) = \frac{1}{\lambda_n}(f * \varphi_n)$, i. e. that either $f * \varphi_n = 0$, or $f * \varphi_n$ is an eigenfunction of L with the same eigenvalue $1/\lambda_n$. Since the eigensubspace corresponding to each eigenvalue of the problem (34) is one-dimensional, then there exist constants $\chi_n = \chi_n(f)$, such that (41) holds. Obviously, $\chi_n(f)$ are linear continuous functionals in $\mathscr{C}([0, 1])$. Using (41) several times, we get $(f * g) * \varphi_n = \chi_n(f * g)\varphi_n = f * (g * \varphi_n) = f * (\chi_n(g)\varphi_n) = \chi_n(g)(f * \varphi_n) = \chi_n(g)\chi_n(f)\varphi_n$, whence (42) follows immediately. Now we shall show that $\varphi_n * \varphi_n \neq 0$ for each $n = 1, 2, \ldots$, i. e. that $\chi_n(\varphi_n) \neq 0$ for $n = 1, 2, \ldots$. Indeed, let us assume the contrary, i. e. that there exists an n_0 with $\varphi_{n_0} * \varphi_{n_0} = 0$. Since from (37) it follows that $\varphi_m * \varphi_{n_0} = 0$ for each $m \neq n_0$, then $f * \varphi_{n_0} = 0$, for each $n = 1, 2, \ldots$. If f is an arbitrary finite linear combination of the eigenfunctions φ_n, then $f*\varphi_{n_0}=0$. Using the well-known denseness in $\mathscr{C}([0,1])$ of these linear combinations with respect to the integral norm $\|\ \|_1$, we get $f*\varphi_{n_0}=0$ for each $f\in\mathscr{C}([0,1])$. Since the function $u(t)$, which represents the right inverse operator L of D in (17) is not a divisor of zero of this convolution, then it follows that $\varphi_{n_0}=0$. But this contradicts the assumption that φ_{n_0} is an eigenfunction. From (42) follows $\chi_n(\varphi_m)=0$ for $m\neq n$. □

Lemma 5. *For each $f\in\mathscr{C}([0,1])$ the identities*

$$f*\varphi_n=T_n\{f\}\frac{\chi_n(\varphi_n)\varphi_n}{T_n\{\varphi_n\}},\quad n=1,2,\dots, \tag{43}$$

hold, with χ_n defined as in the previous lemma.

Proof. Identities (43) can easily be verified for the eigenfunction $f=\varphi_m$ $m=1,2,\dots$. Indeed, if $m\neq n$, then from (41) it follows that if $f=\varphi_m$, then $T_n(f)=0$ and $f*\varphi_n=0$, and (43) holds. If $m=n$, then from (41) it follows that (43) is fulfilled. The identity (43) can be proved in the general case by approximation with respect to the integral norm, in the same way as the approximation was used in the previous lemma. □

Theorem 3.3.4. *If $f*g$ is (17), then*

$$T_n\{f*g\}=\frac{1}{\lambda_nT_n\{u\}}T_n\{f\}\,T_n\{g\},\quad n=1.\ 2,\dots, \tag{44}$$

for every $f,\ g\in\mathscr{C}([0,1])$, i. e. the operation $f*g$ is a convolution of the finite Sturm-Liouville integral transformation.

Proof. By Lemma 5 it is clear that

$$\chi_n(f)=\frac{T_n\{f\}\chi_n\{\varphi_n\}}{T_n\{\varphi_n\}}.$$

Now, using the multiplicativity of χ_n with respect to the convolution $*$, we get at once

$$\frac{T_n(f)\chi_n(\varphi_n)}{T_n(\varphi_n)}\cdot\frac{T_n(g)\chi_n(\varphi_n)}{T_n(\varphi_n)}=\frac{T_n(f*g)\chi_n(\varphi_n)}{T_n(\varphi_n)}$$

i. e.

$$T_n(f*g)=\frac{T_n(f)T_n(g)\chi_n(\varphi_n)}{T_n(\varphi_n)}.$$

The still unknown constants can be determined from the representation $Lf=u*f$. Indeed, $u*\varphi_n=L\varphi_n=\frac{1}{\lambda_n}\varphi_n$ and

$$T_n\{u*\varphi_n\}=\frac{1}{\lambda_n}T_n\{\varphi_n\},\quad T_n\{u*\varphi_n\}=\frac{T_n\{u\}T_n\{\varphi_n\}\chi_n(\varphi_n)}{T_n\{\varphi_n\}}.$$

Hence $\chi_n(\varphi_n)=\frac{T_n\{\varphi_n\}}{\lambda_nT_n\{u\}}$, and (44) is proved. □

Quite similar is the result for the corresponding Sturm-Liouville finite integral transformation in the case $\beta=0$. Thus the problem to find explicit convolutions for the finite Sturm-Liouville integral transformations is solved.

3.3.3. Convolutions of the finite Bessel integral transformations. The considerations of the previous subsection in general are not applicable to a singular operator D. A classical example is the Bessel differential operator

$$B_\nu = \frac{d^2}{dt^2} + \frac{1}{t}\frac{d}{dt} - \frac{\nu^2}{t^2} = t^{-\nu-1}\frac{d}{dt}t^{2\nu+1}\frac{d}{dt}t^{-\nu}, \quad \nu \geqq 0,$$

considered in the finite interval $[0, 1]$. For it one cannot consider arbitrary eigenvalue problems, as those of Sturm-Liouville. In 3.2.2 we have considered a right inverse of B_ν, defined with an initial condition in the singular point $t=0$. The definition of this right inverse operator, which in this subsection will be denoted by $L_{\nu,0}$, is the following. We consider the space $\mathscr{C}_{\nu-2}$ of the functions, defined for $0<t\leqq 1$ and admitting a representation of the form $f(t)=t^p\tilde{f}(t)$ with $p>\nu-2$ and a continuous function $\tilde{f}(t)$ in $0\leqq t\leqq 1$. If $f\in\mathscr{C}_{\nu-2}$, then we define the function $y=L_{\nu,0}f$ as the solution of the initial value problem

$$B_\nu y=f, \quad \lim_{t\to+0} t^{-\nu}y(t)=0.$$

Then $L_{\nu,0}$, as in 3.2.2, has a representation of the form

$$L_{\nu,0}f(t)=\frac{t^2}{4}\int_0^1\int_0^1 t_1^{-\frac{\nu}{2}}t_2^{\frac{\nu}{2}} f(t\sqrt{t_1t_2})\,dt_1\,dt_2. \tag{45}$$

We have $L_{\nu,0}\colon \mathscr{C}_{\nu-2}\to\mathscr{C}^2_\nu$, where by $\mathscr{C}^2_\nu$ we have denoted the space of the functions of the form $f(t)=t^p\tilde{f}(t)$ with $p>\nu$ and with $\tilde{f}(t)\in\mathscr{C}^2([0,1])$. Further we consider the most important case of an integer $\nu=n\geqq 0$ only. The general case can be treated in the same way, but using operators of fractional differentiation.

At first, we consider an arbitrary right inverse operator L_n of B_n in C_{n-2}, defined by a linear functional Φ in the space $\mathscr{C}_{n-2}$ of the functions of the form $f(t)=t^p\tilde{f}(t)$ with $p>n-2$ and with $\tilde{f}(t)\in\mathscr{C}([0,1])$, in such a way that $y=L_nf$ for $f\in\mathscr{C}_{n-2}$ to be the solution of the boundary value problem

$$B_ny=f, \quad \Phi(y)=0. \tag{46}$$

In order problem (46) to be solvable in the whole space $\mathscr{C}_{n-2}$, we should assume $\Phi(\{x^n\})\neq 0$. Then, without any loss of generality, we may assume

$$\Phi(\{x^n\})=1.$$

Then, the right inverse L_n of B_n, we are interested in, is well defined and it has a representation of the form

$$L_nf=L_{n,0}f-\Phi(L_{n,0}f)\{t^n\}. \tag{47}$$

Our aim is to show that (47) has a non-trivial convolution in $\mathscr{C}_{n-2}$, and to find an explicit representation of such a convolution. Here we again rely on the similarity method. In 3.2.2 we have shown that the transformation

$$T_nf(t)=\frac{t^{n+1}}{\Gamma(n+1/2)}\int_0^1(1-\tau)^{n-1/2}\tau^{-n/2}f(t\sqrt{\tau})\,d\tau, \tag{48}$$

referred to as the Sonine transform, is a similarity of $L_{n,0}$ to l^2, i. e.

(49) $$T_n L_{n,0} = l^2 T_n.$$

But, unfortunately, T_n is not a similarity of L_n to a right inverse operator of d^2/dt^2, except in the case $n=0$, since

$$T_n\{t^n\} = \frac{t^{2n+1}}{\Gamma(n+3/2)}.$$

The Sonine transform T_n can be so modified as to become such a similarity.

Definition 1. The *modified Sonine transform* $\tilde{T}_n$ is said to be the transformation

(50) $$\tilde{T}_n f(t) = (T_n f(t))^{(2n)} = \left(\frac{d}{dt}\right)^{2n} \left\{\frac{t^{n+1}}{\Gamma(n+1/2)} \int_0^1 (1-\tau)^{n-1/2} \tau^{-n/2} f(t\sqrt{\tau})\, d\tau\right\}.$$

It is easily seen that the image $\tilde{T}_n(\mathscr{C}_{n-2})$ of $\mathscr{C}_{n-2}$ under $\tilde{T}_n$ is a subspace of $\mathscr{C}_{-1}$, and

$$\tilde{T}_n\{t^n\} = \frac{2^{2n+1} n!}{\sqrt{\pi}} t.$$

Lemma 6. *The modified Sonine transform $\tilde{T}_n$ is a similarity of the right inverse L_n of B_n, defined by (47) to the right inverse operator of d^2/dt^2, defined by*

(51) $$\tilde{L}_n f(t) = l^2 f(t) - \tilde{\Phi}(l^2 f)\, t,$$

where $\tilde{\Phi} = \frac{2^{2n+1} n!}{\sqrt{\pi}} \Phi \circ T_n^{-1}$ *is a linear functional in* $\mathscr{C}_{-1}$.

Proof. The only thing we should show is to establish that the functional $\Phi \circ \tilde{T}_n^{-1}$ is in fact defined on the whole space $\mathscr{C}_{-1}$. To this end we shall use the inversion formula (41). We get the following inversion formula for $\tilde{T}_n$:

(52) $$\tilde{T}_n^{-1} f(t) = \frac{t^n}{2^{n+1}\sqrt{\pi}} \left(\frac{1}{t}\frac{d}{dt}\right)^{n+1} t \int_0^1 (1-\tau)^{-1/2} (l^{2n} f)(t\sqrt{\tau})\, d\tau.$$

Since the function $l^{2n} f$ is at least $2n$ times differentiable, then the inverse transformation $\tilde{T}_n^{-1}$ is defined in the whole space $\mathscr{C}_{-1}$, and $\tilde{T}_n^{-1}(\mathscr{C}_{-1}) \subset \mathscr{C}_{n-2}$. Hence, $\tilde{L}_n$ is a well defined operator in $\mathscr{C}_{-1}$, and $\tilde{T}_n L_n = \tilde{L}_n \tilde{T}_n$, and thus all is proved.

Theorem 3.3.5. *The operation*

(53) $$(f * g)(t) = -\Phi_x \Big\{ (L^n_{n,0})_x (L^n_{n,0})_t (\tilde{T}_n^{-1})_x (\tilde{T}_n^{-1})_t$$

$$\Big[\frac{1}{2} \int_t^x (\tilde{T}_n f)^{(2n)}(x+t-\tau)(\tilde{T}_n g)^{(2n)}(\tau)\, d\tau$$

$$-\frac{1}{2}\int_{-t}^{x}(\tilde{T}_n f)^{(2n)}(|x-t-\tau|)(\tilde{T}_n g)^{(2n)}(|\tau|)\operatorname{sgn}(x-t-\tau)\,\tau\,d\tau\,\Big]\Big\}$$

is a convolution of the operator L_n in $\mathscr{C}_{n-2}$, such that

$$L_n f=\left\{\frac{\sqrt{\pi}}{2^{2n+1}n!}t^n\right\}*f. \tag{54}$$

Proof. By Theorem 1.3.6, the operation $f*g=\tilde{T}_n^{-1}[(\tilde{T}_n f)\,\tilde{*}\,(\tilde{T}_n g)]$, where the operation $\tilde{*}$ is given by $(\tilde{f}*g)(t)=-(\Phi\circ l)_x\Big\{\frac{1}{2}\int_t^x f(x+t-\tau)g(\tau)d\tau$

$-\frac{1}{2}\int_{-t}^{x} f(|x-t-\tau|)g(|\tau|)\operatorname{sgn}(x-t-\tau)\,\tau d\tau\Big\}$ is a convolution of L_n in $\mathscr{C}_{n-2}$.

But (53) is only the explicit form of this convolution. The representation (54) is a consequence of the representation $\tilde{L}_n f=\{t\}*f$, established by Theorem 3.1.3.

Example. Let us consider the simplest but important special case $n=0$. For the sake of simplicity, we confine ourselves to the case $\Phi(f)=f(1)$. For $n=0$ the Sonine transform T_0, which now has the form

$$T_0 f(t)=\frac{2}{\sqrt{\pi}}\int_0^t\frac{\tau f(\tau)}{\sqrt{t^2-\tau^2}}\,d\tau, \tag{55}$$

is a similarity of L_0 to the right inverse

$$\tilde{L}_0 f(t)=t^2 f(t)-\frac{2t}{\pi}\int_0^1 \arccos\tau\, f(\tau)\,d\tau \tag{56}$$

of d^2/dt^2. Then convolution (53) takes the simpler form

$$(f*g)(t)=-\frac{1}{2\sqrt{\pi}}\int_0^1 (T_0^{-1})_t\Big\{\int_t^x (T_0 f)(x+t-\tau)(T_0 g)(\tau)d\tau \tag{57}$$

$$-\int_{-t}^{x}(T_0 f)(|x-t-\tau|)(T_0 g)(|\tau|)\operatorname{sgn}(x-t-\tau)\,\tau d\tau\Big\}\frac{dx}{\sqrt{1-x^2}},$$

where

$$T_0^{-1}f(t)=\frac{1}{\sqrt{\pi}\,t}\,\frac{d}{dt}\int_0^t\frac{u f(u)du}{\sqrt{t^2-u^2}}\,.$$

The convolutions (53) are of interest both on their own right and as containing a definite solution of the problem for finding convolutions of the finite Bessel transforms (Churchill [25], p. 424).

Definition 2. The *finite Hankel integral transform* with a parameter h is said to be the correspondence

$$\mathfrak{H}_{n,j}^{(h)}\{f\}=\int_0^1 f(t)t\,J_n(\mu_j t)dt,\quad j=1,2,\ldots, \tag{58}$$

where μ_j, $j=1, 2,\ldots$, are the positive zeros of the function $hJ_n(t)+tJ_n'(t)$.

In other words, this is the map of a function $f\in\mathscr{C}_{n-2}$ into its Fourier coefficients sequence on the system $\{J_n(\mu_j t)\}_{j=1}^{\infty}$ using the inner product $(f, g) = \int_0^1 tf(t)g(t)dt$.

The basic operational property of the finite Hankel integral transform (58) is given by the identities

$$\mathfrak{H}_{n,j}^{(h)}\{B_n f\}=-\mu_j^2\,\mathfrak{H}_{n,j}^{(h)}\{f\}+J_n(\mu_j)\,[hf(1)+f'(1)], \tag{59}$$

$j=1, 2,\ldots$ (see Churchill [25], p. 425). This shows that it is connected with the singular eigenvalue problem

$$B_n y-\lambda y=0,\quad hy(1)+y'(1)=0 \tag{60}$$

for the Bessel differential operator B_n in $[0, 1]$. It is near to guess that (58) is connected with the right inverse operator of B_n, defined by

$$B_n L_n f(t)=f(t),\quad h(L_n f)(1)+(L_n f)'(1)=0 \tag{61}$$

for $f\in\mathscr{C}_{n-2}$.

Lemma 7. *The right inverse L_n of B_n, defined by* (61), *has a representation of the form*

$$L_n f(t)=L_{n,0}f(t)+\frac{t^n}{h+n}\left\{\frac{2n-1}{2n}\int_0^1 \tau^{n+1} f(\tau)d\tau+\frac{1}{2n}\int_0^1 \tau^{1-n}f(\tau)\,d\tau\right\} \tag{62}$$

in $\mathscr{C}_{n-2}$. A convolution of L_n in $\mathscr{C}_{n-2}$ can be obtained from (53) *by the substitution $\Phi(f)=[hf(1)+f'(1)]/(n+h)$. Then $L_n f=\left\{\frac{\sqrt{\pi}}{2^{2n+1}\,n!}t^n\right\}*f$.*

The proof is immediate.

Now we shall show that this convolution $f*g$ of L_n in $\mathscr{C}_{n-2}$ is a convolution of the finite Hankel integral transform (58) too in the sense of the following

Definition 3. An operation $f*g$ in $\mathscr{C}_{n-2}$ is said to be a convolution of the finite Hankel integral transform $\mathfrak{H}_{n,j}^{(h)}$ iff relations

$$\mathfrak{H}_{n,j}^{(h)}\{f*g\}=\omega_{n,j}^{(h)}\,\mathfrak{H}_{n,j}^{(h)}\{f\}\,\mathfrak{H}_{n,j}^{(h)}\{g\}, \tag{63}$$

with non-zero constants $\omega_{n,j}^{(h)}$, $j=1, 2,\ldots$, hold.

Theorem 3.3.6. *The operation* (53) *with* $\Phi(f)=[hf(1)+f'(1)]/(h+n)$ *is a convolution of the finite Hankel integral transform* (58), *such that*

$$(64)\qquad \mathfrak{H}_{n,j}^{(h)}\{f*g\}=-\frac{\sqrt{\pi}}{2^{2n+1}\,n!\,(h+n)J_n(\mu_j)}\,\mathfrak{H}_{n,j}^{(h)}\{f\}\,\mathfrak{H}_{n,j}^{(h)}\{g\},$$

$j=1, 2,\dots$.

Proof. Relation (64) can be proved just in the same way as the proof of Theorem 3.3.4. To relation (44) corresponds the relation

$$(65)\qquad \mathfrak{H}_{n,j}^{(h)}\{f*g\}=-\frac{1}{\mu_j^2\mathfrak{H}_{n,j}^{(h)}\{u\}}\,\mathfrak{H}_{n,j}^{(h)}\{f\}\mathfrak{H}_{n,j}^{(h)}\{g\},$$

where $u(t)=\dfrac{2^{2n+1}n!}{\sqrt{\pi}}\,t^n$ is the function representing L_n in convolution (58). It can easily be seen that

$$\mathfrak{H}_{n,j}^{(h)}\{t^n\}=\frac{J_n(\mu_j)\,(h+n)}{\mu_j^2},$$

and this together with (65) gives the exact values of the constants $\omega_{n,j}^{(h)}$ in (63) for the convolution considered. □

Thus the convolution problem for the finite Hankel integral transforms, posed by Churchill [25], p. 424, is solved.

3.4. APPLICATIONS OF CONVOLUTIONS TO NON-LOCAL BOUNDARY VALUE PROBLEMS

The natural domain of applicability of the convolutions studied in this chapter are some classes of boundary value problems — both local and non-local — for linear differential operators of the second order. The study of such boundary value problems is interesting both from purely mathematical point of view and for its application to linear non-local problems of mathematical physics. Along with the convolutions for boundary value problems for linear differential operators of the second order, the convolutions for boundary value problems for the differentiation operator, studied in Chapter 2, can be used too. Since the object of the classical problems of mathematical physics are boundary value problems for partial differential operators, such convolutions should be extended in spaces of functions of several variables. This can be accomplished in a rather standard manner (see [46]), if we restrict our considerations to rectangular domains. Then the convolutions needed for treating such boundary value problems are in fact tensor products of one-dimensional convolutions. Nevertheless, there are some cases of problems of mathematical physics in which one convolution suffices. In such cases we can look on the other variables as on parameters. This is a well known approach in using integral transforms for solving such boundary value problems.

In this section two kinds of problems are considered. First, in more detail the eigenexpansions of some spectral problems for the square of differentiation are studied. This is done by using the convolutional approach. The results obtained are used for explicit Duhamel representations of the solutions of Bitsadze-Samarskiy and Samarskiy-Ionkin problems. This is an illustration of the use of the convolutional method in mathematical physics.

3.4.1. Eigen expansions for non-local spectral problems for the square of differentiation. In sect. 3.1 the general spectral problem

(1) $$y''+\lambda^2 y=0,\quad y(0)=0,\quad \Phi(y)=0$$

with a non-zero linear functional Φ in $\mathscr{C}^1([0,1])$ was considered. There the inessential restriction $\Phi\{x\}=1$ was imposed, i. e. that $\lambda=0$ is not an eigenvalue of problem (1). In the same section it was shown that the operation

(2) $$(f*g)(x)=-(\Phi\circ l)_\xi\Big\{\frac{1}{2}\int_x^\xi f(\xi+x-\eta)\,g(\eta)d\eta$$

$$-\frac{1}{2}\int_{-x}^{\xi} f(|\xi-x-\eta|)\,g(|\eta|)\operatorname{sgn}(\xi-x-\eta)\,\eta d\eta\Big\}$$

is a convolution of the right inverse operator $Lf(x)=l^2f(x)-x\Phi(l^2f)$ of d^2/dx^2, determined by $(Lf)(0)=0$ and $\Phi(Lf)=0$. It can be represented as the convolutional operator

(3) $$Lf(x)=\{x\}*f.$$

It is easy to see that each eigenvalue of (1) is a zero of the entire function of exponential type

(4) $$E(\lambda)=\Phi_\xi\Big\{\frac{\sin\lambda\xi}{\lambda}\Big\},$$

and conversely. The multiplicity of a zero λ_n of $E(\lambda)$ determines the dimension of the corresponding eigenspace of λ_n. If $\mu_n>1$, then along with $\sin\lambda_n x$, there are μ_n-1 associated functions in this eigenspace $\mathfrak{X}_{\lambda_n;\mu_n}$, and it is the span of them. Then

(5) $$\mathfrak{X}_{\lambda_n;\mu_n}=\ker\,(I+\lambda_n^2L)^{\mu_n}\quad .$$

According to Theorem 1.3.5, there is an element $\varphi_n\in\mathfrak{X}_{\lambda_n;\mu_n}$ such that the operator $P_nf=\varphi_n*f$ is a projector of $\mathscr{C}([0,1])$ onto $\mathfrak{X}_{\lambda_n;\mu_n}$. We can write down this element explicitly.

Theorem 3.4.1. *Let λ_n be an eigenvalue of spectral problem* (1) *and let C_n be a contour containing λ_n and no other zero of $E(\lambda)$ inside. Then the operator*

(6) $$P_nf=f*\varphi_n$$

with

(7) $$\varphi_n(x)=\frac{1}{\pi i}\int_{C_n}\frac{\sin\lambda x}{E(\lambda)}d\lambda$$

is a projector of $\mathscr{C}([0,1])$ onto $\mathfrak{X}_{\lambda_n;\mu_n}$.

Proof. It is easy to verify that $\varphi_n \in \mathfrak{X}_{\lambda_n;\mu_n}$. It remains to show that $\varphi_n * \varphi_n = \varphi_n$. To this end we take contours C_n' and C_n'' such that C_n'' lies inside C_n'. Then, expressing $\varphi_n(x)$ once as an integral on C_n', and then as integral on C_n'', we get

$$(\varphi_n * \varphi_n)(x) = \frac{1}{(\pi i)^2} \iint\limits_{C_n' \times C_n''} \frac{\{\sin \lambda x\} * \{\sin \mu x\}}{E(\lambda)E(\mu)} \, d\lambda \, d\mu$$

$$= \frac{1}{(\pi i)^2} \iint\limits_{C_n' \times C_n''} \frac{E(\mu)\mu \sin \lambda x - E(\lambda)\lambda \sin \mu x}{(\mu^2 - \lambda^2)E(\lambda)E(\mu)} \, d\lambda \, d\mu$$

$$= \frac{1}{\pi i} \int\limits_{C_n'} \frac{\sin \lambda x}{E(\lambda)} \left[\frac{1}{\pi i} \int\limits_{C_n''} \frac{\mu d\mu}{\mu^2 - \lambda^2} \right] d\lambda - \frac{1}{\pi i} \int\limits_{C_n''} \frac{\sin \mu x}{E(\mu)} \left[\frac{1}{\pi i} \int\limits_{C_n'} \frac{\lambda d\lambda}{\mu^2 - \lambda^2} \right] d\mu = \varphi_n(x),$$

where we use the evident formulas

$$\frac{1}{\pi i} \int\limits_{C_n''} \frac{\mu d\mu}{\mu^2 - \lambda^2} = 0 \quad \text{and} \quad \frac{1}{\pi i} \int\limits_{C_n'} \frac{\lambda d\lambda}{\mu^2 - \lambda^2} = -1. \; \square$$

Corollary. If λ_n is a simple zero of $E(\lambda)$, then $\varphi_n(x) = 2 \sin \lambda_n x / E'(\lambda_n)$ and

$$P_n f(x) = -\frac{2}{\lambda_n E'(\lambda_n)} \Phi_\xi \left\{ \int\limits_0^\xi \sin \lambda_n(\xi - \eta) f(\eta) d\eta \right\} \sin \lambda_n x. \tag{8}$$

As in Chapter 2, sect. 2.2, we can introduce a formal eigenexpansion for problem (1).

Definition 1. Let $f(x) \in \mathscr{C}([0, 1])$ and let $\lambda_1, \lambda_2, \ldots$ be the eigenvalues of (1). The *formal spectral expansion* of $f(x)$ for eigenvalue problem (1) is said to be the correspondence

$$f(x) \sim \sum_{n=1}^{\infty} (f * \varphi_n)(x). \tag{9}$$

It is not supposed the series in (9) to be convergent, but nevertheless this formal expansion is of considerable interest, if it has the uniqueness property, i. e. when from $(f * \varphi_n)(x) = 0$ for $n = 1, 2, 3, \ldots$ it follows $f(x) = 0$. Here we shall consider two non-local eigenvalue problems (1) which have the uniqueness property.

1. *The Bitsadze-Samarskiy spectral problem* [5]

Let us consider the eigenvalue problem

$$y'' + \lambda^2 y = 0, \quad y(0) = 0, \quad y(1) - y(1/2) = 0. \tag{10}$$

The functional $\Phi(f)$ should be taken in the form $\Phi(f) = 2[f(1) - f(1/2)]$. There are two series of eigenvalues: $\alpha_n = 2n\pi$, $n = 1, 2, \ldots$, and $\beta_m = 2\pi/3 + 4m\pi$, $m = 0, \pm 1, \pm 2, \ldots$. In this case formal expansion (9) takes the form

$$(11)\qquad f(x)\sim\sum_{n=1}^{\infty} a_n \sin 2n\pi x+\sum_{m=-\infty}^{\infty} b_m \sin(2\pi/3+4m\pi)\,x$$

with

$$a_n=\frac{4}{2-(-1)^n}\Big[\int_0^1 \sin 2n\pi\xi f(\xi)d\xi-(-1)^n\int_0^{1/2}\sin 2n\pi\xi\, f(\xi)d\xi\Big]$$

and

$$b_m=\frac{4}{3}\Big[\int_0^1 \sin\Big(\frac{2\pi}{3}+4m\pi\Big)(1-\xi)f(\xi)d\xi-\int_0^{1/2}\sin\Big(\frac{2\pi}{3}+4m\pi\Big)\Big(\frac{1}{2}-\xi\Big)f(\xi)\,d\xi\Big].$$

Lemma 1. *Expansion* (11) *for the function* $f(x)=L\{x\}=x^3/6-7x/24$ *represents it and is uniformly convergent.*

Proof. Computing the coefficients in (11), we should prove the identity

$$(12)\qquad \frac{x^3}{6}-\frac{7x}{24}=\frac{1}{4\pi^3}\Big[\sum_{n=1}^{\infty}\frac{\sin 2n\pi x}{n^3[2-(-1)^n]}-18\sum_{m=-\infty}^{\infty}\frac{\sin(2\pi/3+4m\pi)x}{(1+6m)^3}\Big].$$

We use the known identity

$$(13)\qquad \sum_{m=-\infty}^{\infty}\frac{\sin(m-a)\theta}{m-a}=\pi,$$

valid for $0<\theta<2\pi$ and for non-integer a (see T. Bromwich. *An Introduction to the Theory of Infinite Series.* Cambridge, 1926, p. 371). From (13) the following two identities can be obtained:

$$\sum_{m=-\infty}^{\infty}\frac{\cos(2\pi/3+4m\pi)x}{(2\pi/3+4m\pi)^2}=\frac{1-x}{4}$$

and

$$\sum_{m=-\infty}^{\infty}\frac{\sin(2\pi/3+4m\pi)x}{(2\pi/3+4m\pi)^3}=\frac{x(2-x)}{8}\text{ for }0\leq x\leq\frac{1}{2}.$$

From the elementary theory of Fourier series the identities

$$\sum_{n=1}^{\infty}\frac{\sin 4n\pi x}{(4n\pi)^3}=\frac{x^3}{12}-\frac{x^2}{16}+\frac{x}{96}$$

and

$$\sum_{n=1}^{\infty}\frac{\sin 2(2n-1)\pi x}{[2(2n-1)\pi]^3}=\frac{x(1-2x)}{32},$$

valid for $0 \leq x \leq 1/2$, can be used. By means of these four identities, it can easily be shown that (12) is valid in $0 \leq x \leq 1/2$. For $1/2 \leq x \leq 1$ the validity of (12) can be shown in the same way, but now using the identity

$$\sum_{m=-\infty}^{\infty} \frac{\cos(m+a)\theta}{m+a} = \frac{\pi}{\operatorname{tg} \pi a}, \quad 0<\theta<2\pi, \quad a \text{ non-integer}$$

(see T. Bromwich, l. c., p. 371, ex. 5). □

Lemma 2. *Let* $f \in \mathscr{C}^4([0, 1])$ *and* $f(0)=f''(0)=0$, $f(1)-f(1/2)=f''(1)-f''(1/2)=0$. *Then expansion* (11) *gives a uniformly convergent series, converging to* $f(x)$ *in* $[0, 1]$.

Proof. From the generalized Taylor formula (24) of 1.3.4 we have

$$f(x)=L^2 f^{(4)}(x)=(L\{x\}) * f^{(4)}(x)=(x^3/6-7x/24) * f^{(4)}(x).$$

Then, by convolutional multiplication of (12) by $f^{(4)}(x)$, we get expansion (11) of $f(x)$. It is uniformly convergent and represents $f(x)$ in $[0, 1]$. □

Theorem 3.4.2. (*Uniqueness property of* (11)). *Let f be an arbitrary function of* $\mathscr{C}([0, 1])$. *If all the coefficients a_n and b_m in* (11) *are* 0, *then* $f(x) \equiv 0$.

Proof. Let $a_n=0$, $n=1, 2, \ldots$, $b_m=0$, $m=0, \pm 1, \pm 2, \ldots$. Then the corresponding coefficients $\tilde{a}_n$ and $\tilde{b}_m$ of the function $\tilde{f}(x)=L^2 f(x)$ are

$$\tilde{a}_n=\frac{1}{(2n\pi)^4} a_n=0, \quad \tilde{b}_m=\frac{1}{\left(\frac{2\pi}{3}+4\pi m\right)^4} b_m=0.$$

Since $\tilde{f}(x)$ satisfies the hypothesis of Lemma 3, then $\tilde{f}(x) \equiv 0$. Applying d^2/dx^2 to the last identity, we get $f(x) \equiv 0$. □

Now we return again to problem (1). There are cases in which a simpler convolution than (2) can be found in $\mathscr{C}([0, 1])$. This is the case when the functional Φ is a smoothing one.

Theorem 3.4.3. *Let* $\Phi=\Psi \circ l$ *with a linear functional* Ψ *in* $\mathscr{C}([0, 1])$, *and* $\Psi_\xi\{\xi^2/2\}=\Phi_\xi\{\xi\}=1$. *Then the operation*

$$(14) \qquad (f \tilde{*} g)(x)=f(x)\Psi(lg)+g(x)\Psi(lf)-\Psi\{1\}\int_0^x f(x-\xi)\, g(\xi) d\xi$$

$$-\Psi_\xi\left\{\frac{1}{2}\int_x^\xi f(\xi+x-\eta) g(\eta) d\eta-\frac{1}{2}\int_{-x}^\xi f(|\xi-x-\eta|)\, g(|\eta|) \operatorname{sgn}(\xi-x-\eta)\eta d\eta\right\}$$

is a convolution of the right inverse L of d^2/dx^2, *determined by* (1). *It has the function* $\{x\}$ *as unit, i. e.*

$$(15) \qquad \{x\} \tilde{*} f=f \quad \text{for all} \quad f \in \mathscr{C}([0, 1]).$$

Proof. In order to show that (14) is a convolution of L in $\mathscr{C}([0, 1])$, we can rely on the identity

$$(f \tilde{*} g)(x) = \frac{d}{dx^2}[(f * g)(x)].$$

Then the associativity relation $(f \tilde{*} g) \tilde{*} h = f \tilde{*} (g \tilde{*} h)$ follows from that for $*$, using the fact that $(f * g)(0)=0$ and $\Phi(f * g)=0$ for all $f, g \in \mathscr{C}([0,1)]$. Identity (15) can be verified directly. Since $Lf=\{x\}*f$, then $Lf=(L\{x\})\tilde{*}f=\{x^3/6-x\Psi_\xi(\xi^2/2)\}*f$ and hence $\tilde{*}$ is a convolution of L in $\mathscr{C}([0, 1])$. □

Lemma 3. *Let the functional* Φ *in* (1) *be the same as in Theorem* 3.4.3. *Then, if* λ_n *is an eigenvalue of* (1), *the operator*

(16) $$P_n f(x) = (f \tilde{*} \tilde{\varphi}_n)(x)$$

with

(17) $$\tilde{\varphi}_n(x) = -\frac{1}{\pi i}\int\limits_{C_n} \frac{\sin \lambda x}{\lambda^2 E(\lambda)}\, d\lambda$$

is a projector of $\mathscr{C}([0, 1])$ *onto the eigenspace* $\mathfrak{X}_{\lambda_n;\mu_n}$ *of* λ_n.

Proof. Since $\tilde{\varphi}_n = L\varphi_n$, then using Theorem 3.4.1, we obtain

$$P_n f = f * \varphi_n = f \tilde{*} \tilde{\varphi}_n. \ \square$$

Using convolution (14), the formal eigenexpansion (9) can be written in the form

(18) $$f(x) \sim \sum_{n=1}^{\infty} (f \tilde{*} \tilde{\varphi}_n)(x).$$

We shall use convolution (14) for a study of a special non-local spectral problem we shall need later.

2. The Samarskiy-Ionkin spectral problem [65, 66].

This is the eigenvalue problem

(19) $$y'' + \lambda^2 y = 0,\ y(0)=0,\ \int\limits_0^1 y(\xi)d\xi = 0.$$

Here $\Phi(f) = 2\int\limits_0^1 f(\xi)\, d\xi$, and

$$E(\lambda) = \frac{2(1-\cos\lambda)}{\lambda^2}.$$

The eigenvalues of (19) are $\lambda_n = 2n\pi$, $n=1, 2, \ldots$, and each of them has multiplicity 2. Hence, the eigenspaces $\mathfrak{X}_{\lambda_n;2}$ are two-dimensional. Then convolution (14) takes the form

(20) $$(f \tilde{*} g)(x) = 2f(x)\int\limits_0^1 g(\xi)d\xi + 2g(x)\int\limits_0^1 f(\xi)d\xi$$

$$-2\int_0^x f(x-\xi)g(\xi)d\xi-\int_x^1 f(1+x-\xi)g(\xi)d\xi$$

$$+\int_{-x}^1 f(|1-x-\xi|)\,g(|\xi|)\,\mathrm{sgn}\,(1-x-\xi)\xi d\xi.$$

Lemma 4. *The convolutional projectors* (16) *for* (19) *have the form*

$$P_nf(x)=\{-2x\cos 2n\pi x\}\tilde{*}f(x) \tag{21}$$

$$=\left[4\int_0^1(1-\xi)f(\xi)\sin 2n\pi\xi d\xi\right]\sin 2n\pi x-\left[4\int_0^1 f(\xi)(1-\cos 2n\pi\xi)d\xi\right]x\cos 2n\pi x.$$

Proof. It is easy to get from (17)

$$\tilde{\varphi}_n(x)=-2x\cos 2n\pi x. \tag{22}$$

Then $P_nf(x)=(\tilde{\varphi}_n\tilde{*}f)(x)$ can be transformed into (21) in a standard manner. □

Theorem 3.4.4. *Let* $f\in\mathscr{C}([0,1])$. *If* $P_nf=0$ *for* $n=1,2,\ldots,$ *then* $f(x)\equiv 0$.

Proof. If $P_nf=0$, then $P_nLf=0$ for $n=1,2,\ldots$. The function Lf satisfies both $(Lf)(0)=0$ and $\int_0^1(Lf)(\xi)d\xi=0$. From (21) we have

$$\int_0^1(Lf)(\xi)\cos 2n\pi\xi d\xi=0\quad\text{and}\quad\int_0^1(1-\xi)(Lf)(\xi)\sin 2n\pi\xi d\xi=0.$$

From the first of them it follows that $Lf(x)$ is odd with respect to the point $x=1/2$, i. e. $(Lf)(x)=-(Lf)(1-x)$. The second of the above equations gives $\int_0^1(Lf)(\xi)\sin 2n\pi\xi d\xi=0$. Hence $Lf(x)$ should be even with respect to the point $x=1/2$. This is possible only for $Lf(x)\equiv 0$. Hence $f(x)\equiv 0$.

Hence spectral development (18) of a function $f(x)\in\mathscr{C}([0,1])$ determines it uniquely, though it may be non-convergent. □

Lemma 5. *Expansion* (18) *for the function* $f(x)=L\{x\}=x^3/6-x/12$ *gives a uniformly convergent series, which represents it in* $[0,1]$.

Proof. The identity

$$\frac{x^3}{6}-\frac{x}{12}=\sum_{n=1}^{\infty}\left\{\frac{2x\cos 2n\pi x}{(2n\pi)^2}-\frac{4\sin 2n\pi x}{(2n\pi)^3}\right\} \tag{23}$$

can be proved by elementary summings of the Fourier series involved. □

Theorem 3.4.5. *Let* $f(x)\in\mathscr{C}^2([0,1])$, $f(0)=0$, *and* $\int_0^1 f(\xi)d\xi=0$. *Then*

$$(24)\qquad f(x)=\sum_{n=1}^{\infty}\Big\{\Big[4\int_0^1(1-\xi)f(\xi)\sin 2n\pi\xi d\xi\Big]\sin 2n\pi x$$
$$-\Big[4\int_0^1 f(\xi)(1-\cos 2n\pi\xi)d\xi\Big]x\cos 2n\pi x\Big\},$$

the series being uniformly convergent on $[0, 1]$.

Proof. Under the hypothesis

$$f(x)=Lf''=\{x^3/6-x/12\}\tilde{*}f''.$$

Then by convolutional multiplication of (23) by $f'(x)$, we get

$$f(x)=\sum_{n=1}^{\infty}(L\tilde{\varphi}_n)\tilde{*}f''=\sum_{n=1}^{\infty}\tilde{\varphi}_n\tilde{*}Lf''=\sum_{n=1}^{\infty}\tilde{\varphi}_n\tilde{*}f$$

which is (24). □

Now let us consider the multiplier problem for the expansion (24) in $\mathscr{C}([0, 1])$.

Definition 2. An operator $M:\mathscr{C}([0,1])\to\mathscr{C}([0,1])$ is said to be a multiplier of the formal eigenexpansion (18), if

$$Mf\sim\sum_{n=1}^{\infty}\mu_n(f\tilde{*}\tilde{\varphi}_n)$$

with numerical multiplier sequence $\mu_1, \mu_2, \ldots, \mu_n, \ldots$.

Using the uniqueness Theorem 3.4.4 it can easily be shown that each multiplier of (18) is a multiplier of convolution (20) too. The converse is not always true. Therefore, each multiplier of the Samarskiy-Ionkin eigenexpansion (18) has the form

$$(25)\qquad Mf=m\tilde{*}f,\quad \text{with } m\in\mathscr{C}([0,1]).$$

It remains only the form of m to be specified.

Theorem 3.4.6. *A linear operator* $M:\mathscr{C}([0,1])\to\mathscr{C}([0,1])$ *is a multiplier of the Samarskiy-Ionkin eigenexpansion* (24) *iff it has the form* (25) *with*

$$m(x)=xh(|1-2x|)$$

where h *is an arbitrary function from* $\mathscr{C}([0,1])$.

Proof. It is easy to see that in order (25) to be multiplier of (24), m should have expansion of the form

$$m(x)\sim\sum_{n=1}^{\infty}\mu_n\tilde{\varphi}_n(x).$$

Hence

$$\int_0^1 (1-\xi)\, m(\xi) \sin 2n\pi\xi d\xi = 0, \quad n=1, 2, \dots .$$

This means that $(1-x)m(x)$ should be an even function with respect to the point $x=1/2$, i. e.

$$(1-x)\, m(x) = xm\,(1-x).$$

Hence the function $m(x)/x$ should be continuous on $[0, 1]$ and even with respect to the point $x=1/2$. The general form of the even functions with respect to $x=1/2$ is $h(|1-2x|)$ with $h(x) \in \mathscr{C}([0, 1])$. □

It deserves to be mentioned that the *multiplier problem for the Samarskiy-Ionkin formal expansion* allows an easier solution than for the classical Fourier expansion.

3.4.2. Duhamel-type representations of solutions of non-local boundary value problems for partial differential equations of mathematical physics. Now we shall apply the considerations of the previous section to some problems for equations of mathematical physics. Since our aim here is rather to illustrate the applicability of the convolutions found than to treat elaborately the most general problems, we confine ourselves to only two representative examples.

The simplest and well-known example of a Duhamel representation concerns the boundary value problem

$$(26) \qquad \begin{aligned} &u_t = u_{xx}, \\ &u(0, t) = 0, \quad u(1, t) = f(t); \quad u(x, 0) = 0 \end{aligned}$$

for the heat equation. The solution $u(x, t)$ of (25) can be represented by the Duhamel integral

$$(26') \qquad u(x, t) = \frac{\partial}{\partial t} \int_0^t U(x, t-\tau) f(\tau) d\tau,$$

where $U(x, t)$ is the special solution of (26) with $f(t) \equiv 1$. We observe at once the appearance of the Duhamel convolution in (26′). It is rather natural to seek another boundary value problems for the heat equation, when similar representations hold, but with other convolutions. Especially, we are interested in Duhamel-type representations with respect to the space variable.

Let us consider the general boundary value problem

$$(27) \qquad \begin{aligned} &u_t = u_{xx}, \\ &u(0, t) = 0, \quad \Phi_\xi\{u(\xi, t)\} = 0; \quad u(x, 0) = f(x) \end{aligned}$$

in $D = \{(x, t): 0 < x < 1,\ t > 0\}$ with non-zero linear functional Φ in $\mathscr{C}([0, 1])$. It is not a strong restriction to assume that $\Phi\{x\} = 1$. To this problem the following theorem can be proved.

Theorem 3.4.7 (Dimovski, Mineff [48]). *Let there exist a solution* $U(x,t)$ *of* (27), *with* $f(x)=x^3/6-\Phi_\xi(\xi^3/3)\,x$. *If* $f(x)\in\mathscr{C}^4([0,1]), f(0)=f''(0)=0$, *and* $\Phi(f)=\Phi(f'')=0$, *then the function*

$$u(x,\ t)=\frac{\partial^4}{\partial x^4}\,[U(x,t)\overset{(x)}{*}f(x)], \tag{28}$$

where $\overset{(x)}{*}$ *denotes operation* (2), *is a solution of boundary value problem* (27).

Proof. Under the hypothesis

$$u(x,\ t)=U(x,\ t)\overset{(x)}{*}f^{(4)}(x)$$

and it can directly be verified that $u(x,t)$ is a solution of the heat equation $u_t=u_{xx}$ in D. The boundary value conditions are satisfied since operation (2) has the properties $(f*g)(0)=0$, and $\Phi(f*g)=0$ for arbitrary $f,\ g\in\mathscr{C}([0,\ 1])$. □

Theorem 3.4.7 is only a conditional existence theorem. In each case the existence of the special solution $U(x,\ t)$ for the case $f(x)=x^3/6-x\Phi_\xi(\xi^2/2)$ should be ensured. There easily can be given examples when such a solution does not exist. This is the case especially for some inverse ill-posed problems for the heat equation. The usefulness of representation (28) is in its universality.

Now we shall consider the Samarskiy-Ionkin problem (see [66]) in which the existence of the special solution $U(x,\ t)$ can easily be proved using the considerations of 3.4.1. This problem is announced to be important for plasma physics. It is a mathematical model of diffusion in turbulent plasma [65, 66].

The *Samarskiy-Ionkin problem* is said to be the following non-local boundary value problem for the heat equation:

$$\begin{gathered} u_t=u_{xx},\\ u(0,\ t)=0,\quad \int_0^1 u(\xi,\ t)\,d\xi=0,\\ u(x,\ 0)=f(x). \end{gathered} \tag{29}$$

in the strip $D=\{(x,\ t):0<x<1,\quad t>0\}$.

Let us try to find the solution $U(x,\ t)$ of (24) for $f(x)=L\{x\}=\frac{x^3}{6}-\frac{x}{12}$, where $Lf(x)=l^2f(x)-x\int_0^1(1-\xi)^2f(\xi)\,d\xi$ is the right inverse of d^2/dx^2, determined by $(Lf)(0)=0$ and $\int_0^1(Lf)(\xi)d\xi=0$. To this end we apply the projector operators (21) to the equation $u_t=u_{xx}$ and to the initial condition $u(x,\ 0)=f(x)$. Denoting $u_n(x,\ t)=P_nu(x,\ t)$, we shall have

$$u_n(x,\ t)=A_n(t)(-2x\cos 2\pi nx)+B_n(t)\sin 2\pi nx$$

with unknown functions $A_n(t)$ and $B_n(t)$. It is easy to see that $u_n(x, t)$ should be a solution of the heat equation, such that $u_n(x, 0)=f_n(x)$ with $f_n(x)=P_nf(x)$, given by (21). In our special case $f(x)\equiv L\{x\}$, we have

$$f_n(x)=P_nL\{x\}=LP_n\{x\}=-\frac{1}{4\pi^2n^2}(-2x\cos 2\pi nx)-\frac{1}{2\pi^3n^3}\sin 2\pi nx.$$

For $A_n(t)$ and $B_n(t)$ we get easily the system of ordinary differential equations

$$A_n'(t)=-4\pi^2n^2A_n(t),$$

$$B_n'(t)=-4\pi^2n^2B_n(t)+8\pi nA_n(t)$$

with the initial conditions $A_n(0)=-1/(4\pi^2n^2)$ and $B_n(0)=-1/(2\pi^3n^3)$. We get $A_n(t)=-\exp(-4\pi^2n^2t)/(2\pi n)^2$ and $B_n(t)=-(1+4\pi^2n^2t)\exp(-4\pi^2n^2t)/(2\pi^3n^3)$, $n=1, 2,\dots$. Therefore, if there exists a solution of (25) with $f(x)\equiv L\{x\}$ $=\frac{x^3}{6}-\frac{x}{12}$, the only candidate for such a solution is

$$U(x, t)=\sum_{n=1}^{\infty}\frac{1}{2\pi^3n^3}[\pi nx\cos 2\pi nx-(1+4\pi^2n^2t)\sin 2\pi n x]e^{-4\pi^2n^2t}. \tag{30}$$

The series is uniformly convergent on $\bar{D}$. It is easy to verify that $U(x, t)$ satisfies the heat equation $u_t=u_{xx}$ in $\bar{D}$. It remains only to see that $U(x,t)$ satisfies the boundary value conditions $U(0, t)=0$, $\int_0^1 U(\xi, t)\,d\xi=0$ and the initial value condition $U(x, 0)=x^3/6-x/12$. The boundary value conditions are satisfied since each $u_n(x, t)$ satisfies them. As for the initial value condition, it is satisfied due to (23).

Combining representation (28) with the convolution (20), we can prove the following theorem.

Theorem 3.4.8. *If* $f(x)\in \mathscr{C}^2([0, 1])$, $f(0)=0$ *and* $\int_0^1 f(\xi)d\xi=0$, *then problem* (29) *has a classical solution. It can be represented in the form*

$$u(x, t)=-2\int_0^x \Omega(x-\xi, t)f(\xi)d\xi-\int_x^1 \Omega(1+x-\xi, t)f(\xi)\,d\xi \tag{31}$$

$$+\int_{-x}^1 \Omega(1-x-\xi, t)f(|\xi|)\operatorname{sgn}\xi\, d\xi$$

with

$$\Omega(x, t)=U_{xx}(x, t)=\sum_{n=1}^{\infty}\{-2x\cos 2n\pi x+8\pi nt\sin 2n\pi x\}e^{-4n^2\pi^2t}. \tag{32}$$

P r o o f. The existence of a classical solution of (29) follows from Theorem 3.4.7, since there exists the special solution for $f(x)=x^3/6-x/12$. In D (28) can be represented in the form

$$u(x, t)=U_{xx}(x, t)\tilde{*}f(x).$$

Then, using convolution (20), we can write at once (31), since $\int_0^1 U_{xx}(\xi, t)d\xi=0$.

for $t>0$ and $\int_0^1 f(\xi)d\xi=0$. □

The restriction $f\in\mathscr{C}^2([0, 1])$ can be replaced by the weaker assumption $f\in\mathscr{C}^1([0, 1])$.

The *Bitsadze-Samarskiy problem* (see [5])

$$\begin{gathered} u_{xx}+u_{yy}=0, \\ u(0, y)=0, \quad u(1, y)=u(1/2, y), \\ u(x, 0)=0, \quad u(x, 1)=f(x) \end{gathered} \tag{33}$$

in the square domain $D=\{(x, y): 0<x<1, 0<y<1\}$ is another example of non-local boundary value problem, well suited to the convolutional approach developed here.

L e m m a 6. *The function*

$$U(x, y)=\frac{1}{4\pi^3}\sum_{n=1}^{\infty}\frac{\sin 2n\pi x \,.\, \mathrm{sh}\, 2n\pi y}{[2-(-1)^n]n^3\,\mathrm{sh}\, 2n\pi} \tag{34}$$

$$-\frac{9}{2\pi^3}\sum_{m=-\infty}^{\infty}\frac{\sin(2\pi/3+4m\pi)x \,.\, \mathrm{sh}\,(2\pi/3+4m\pi)y}{(1+6m)^3\,\mathrm{sh}\,(2\pi/3+4m\pi)}$$

is a solution of (33) *for* $f(x)=\frac{x^3}{6}-\frac{7x}{24}$.

P r o o f. It is easy to see that $U(x, t)$ satisfies the Laplace equation in each interior point of the square. The first three boundary value conditions are satisfied in an obvious way. As for the last condition, its satisfaction follows from Lemma 1.

T h e o r e m 3.4.9. *Let* $f(x)\in\mathscr{C}^4([0, 1])$ *and* $f(0)=f''(0)=f(1)-f(1/2)=f''(1)-f''(1/2)=0$. *Then the function*

$$u(x, y)=-\int_{1/2}^{1}d\xi\Big\{\int_x^{\xi}U(x+\xi-\eta, y)f^{(4)}(\eta)d\eta-\int_{-x}^{\xi}U(\xi-x-\eta, y)f^{(4)}(|\eta|)\,\mathrm{sgn}\,\eta d\eta\Big\}, \tag{35}$$

where $U(x, y)$ *is the special solution* (34), *is a classical solution of Bitsadze-Samarskiy boundary value problem* (33) *on the square* $\bar{D}$.

Proof. Since according to Duhamel-type representation (28)

$$u(x, y)=U(x, y) \overset{(x)}{*} f^{(4)}(x),$$

then $u_{xx}+u_{yy}=\{U_{xx}+U_{yy}\} \overset{(x)}{*} f^{(4)}(x)=0$. The first three boundary value conditions in (33) are satisfied in an obvious way. As for the condition $u(x, 1) =f(x)$, its satisfaction follows from Lemma 2.

The Duhamel-type representations (31) and (35) can be used for numerical calculation of the solution at points where we need its values with a prescribed accuracy. To this end some quadrature formulas can be used. An advantage of such a numerical method over the difference methods consists in the possibility to reach greater accuracy and in the stability of the method.

Duhamel-type representation can be proposed for a broad class of linear partial differential equations in rectangular domains. In [46] the class of partial differential equations

$$\sum_{j=1}^{m} P_j(\partial/\partial t_j)u-\sum_{k=1}^{n} Q_k(\partial^2/\partial x_k^2)u=f(x_1,\ldots, x_n; t_1,\ldots,t_m),$$

with polynomials P_j and Q_k is considered in domains $D=[0, 1]^n\times[0, \infty)^m$. The boundary value problems considered there are of the form

$$\chi_{t_j}^{(j)}\{(\partial^l/\partial t_j^l)u\}=f_j^{(l)},\ l=0, 1,\ldots, \deg P_j-1,\ j=1, 2,\ldots, m$$

with respect to the "time variables" $t_1, t_2,\ldots, t_m$, and

$$\left.\frac{\partial^{2s}u}{\partial x_k^{2s}}\right|_{x_k=0}=g_k^{(s)},$$

$$\Phi_{x_k}^{(k)}\{(\partial^{2s}/\partial x_k^{2s})u\}=h_k^{(s)},\quad s=0, 1,\ldots, \deg Q_k-1,\quad k=1, 2,\ldots, n$$

with respect to the "space variables" $x_1, x_2,\ldots, x_n$. Here $\chi^{(1)}, \chi^{(2)},\ldots, \chi^{(m)}$ and $\Phi^{(1)}, \Phi^{(2)},\ldots, \Phi^{(n)}$ are non-zero, but otherwise arbitrary linear functionals on $\mathscr{C}([0, \infty))$ and $\mathscr{C}^1([0, 1])$ correspondingly. $f_j^{(l)}$, $g_k^{(s)}$ and $h_k^{(s)}$ are given boundary value functions. In all these boundary value problems the existence problem can be reduced to the existence of a solution of a simpler problem with simple boundary value functions. It seems that the potentialities of the convolutional method for linear non-local boundary value problems are greater than the examples considered here indicate.

REFERENCES

Berg, L.

[1] Operatorenrechnung I. Algebraische Methoden. Berlin, 1972.

[2] Generalized convolutions. — Math. Nachr., **72**, 1976, 239-245.

Bittner, R.

[3] Operational calculus in linear spaces. — Studia Math., **20**, 1961, 1-18.

[4] Algebraic and analytic properties of solutions of abstract differential equations. — Rozprawy Mathematiczne, **41**, 1964, 1-63.

Bitsadze, Samarskiy [А. В. Бицадзе, А. А. Самарский]

[5] О некоторых простейших обобщениях линейных эллиптических краевых задач.— ДАН СССР, **185**, 4, 1969, 739-741.

Boehme, T. K.

[6] The Mikusiński operators as a topological space. — Amer. J. Math., **98**, 1, 1976, 55-66.

[7] Convolution factoring on the half-line. — Compt. rend. Acad. bulg. Sci., **33**, 5, 1980, 595-598.

Boehme, T. K., Wygant, G.

[8] Generalized functions on the unit circle. — Amer. Math. Monthly, **82**, 1975, 256-261.

Bozhinov [Н. С. Божинов]

[9] Операционни смятания за частни диференциални оператори от първи и втори ред. — В: Математика и математическо образование. Доклади на VII пролетна конференция на СМБ, Сл. бряг, 5—8 април 1978 г. София, 1978, 231-240.

[10] A convolutional approach to the multiplier problem connected with generalized eigenvector expansions. — В: Математика и математическо образование. Доклади на IX пролетна конференция на СМБ, Сл. бряг, 3 — 6 април 1980 г. София, 1980, 9-16.

[11] Differentiation properties and multipliers of Berg—Dimovski convolution for the differentiation operator. — Serdica. Bulg. Math. Publ., **6**, 3, 1980, 219-239.

[12] Classes of multipliers of Berg—Dimovski convolution and real Dirichlet expansions. — Compt. rend Acad. bulg. Sci., **34**, 9, 1981, 1213-1216.

[13] Differentiation properties, multipliers and commutants, related to Dimovski's convolutions for second order differential operator. — Compt. rend. Acad. bulg. Sci., **34**, 8, 1981, 1057-1060.

[14] Representation of commutants and multipliers, connected with Sturm-Liouville expansions. — In: Constructive Function Theory '81. Proc. Internat. Conf. on Constructive Function Theory. Varna, May 31 — June 6, 1981 (in print).

[15] Convolutions and multiplier projections connected with generalized eigenvector expansions. — Compt. rend. Acad. bulg. Sci., **33**, 1, 1980, 31-34.

[16] Convolutions and multipliers connected with generalized eigenvector expansions. — Compt. rend. Acad. bulg. Sci., **33**, 2, 1980, 151-154.

Bozhinov, Dimovski [Н. С. Божинов, И. Х. Димовски]

[17] Operational calculus for general linear differential operator of second order. — Compt. rend. Acad. bulg. Sci., **28**, 6, 1975, 727-730.

[18] Boundary value operational calculi for linear differential operators of second order. — Compt. rend. Acad. bulg. Sci., **31**, 7, 1978, 815-818.

[19] Operational calculi for boundary value problems. — In: Generalized Functions and Operational Calculus. Proc. Conf., Varna, Sept. 29 — Oct. 6, 1975. Sofia, 1979, 49-60.

[20] A representation of the commutant of the initial Sturm—Liouville operator. — Serdica. Bulg. Math. Publ., **6**, 1980, 153-168.

[21] A generalization of a Raichinov's representation formula for the commutant of integration operators. — Compt. rend. Acad. bulg. Sci., **34**, 4, 1981, 477-480.
[22] A convolutional approach to multiplier problem for multiple complex Dirichlet expansions. — Compt. rend. Acad. bulg. Sci., **34**, 11, 1981, 1477-1480.
[23] Convolutions, multipliers and commutants connected with multiple Dirichlet expansions. — Serdica. Bulg. Math. Publ. (in print).

Brichkov, Prudnikov, Shishov [Ю. А. Брычков, А. П. Прудников, В. С. Шишов]
[24] Операционное исчисление. — В: Итоги науки и техники. Матем. анализ, **16**, 1979, 99-148.

Churchill, R. V.
[25] Operational Mathematics. 3rd ed. New York, 1972.

Cohen, P. J.
[26] Factorization in group algebras. — Duke Math. J., **26**, 1959, 199-205.

Delsarte, J.
[27] Le calcul linéaire. Bull. Soc. Math. France, C. R. des sciences, 1937, 42-53. — In: Oeuvres de Jean Delsarte, I. Paris, 1971, 411-422.
[28] Une extension nouvelle de la théorie des fonctions présque-périodique de Bohr.— Acta Math., **69**, 1938, 259-317.

Dimovski, I. H. [И. Х. Димовски]
[29] Единственост на полето на Микусински. — Физ.-мат. списание, **5** (**38**), 1, 1962, 56-59.
[30] Operational calculus for a class of differential operators. — Compt. rend. Acad. bulg. Sci., **19**, 12, 1966, 1111-1114.
[31] On an operational calculus for a differential operator. — Compt. rend. Acad. bulg. Sci., **21**, 6, 1968, 513-516.
[32] Свертка оператора Сонина. — Compt. rend. Acad. bulg. Sci., **21**, 10, 1968, 1005-1008.
[33] An explicit expression for the convolution of the Meijer transformation. — Compt. rend. Acad. bulg. Sci., **26**, 10, 1973, 1293-1296.
[34] Operational calculus for the general linear differential operator of the first order.— Compt. rend. Acad. bulg. Sci., **26**, 12, 1973, 1579-1582.
[35] On an operational calculus for vector-valued functions. — Math. Balkanica, **4**, 1974, 129-135.
[36] Convolutions for the right inverse linear operators of the general linear differential operator of the first order. — Serdica. Bulg. Math. Publ., **2**, 1, 1976, 82-86.
[37] Two new convolutions for linear right inverse operators of d^2/dt^2. — Compt. rend. Acad bulg. Sci., **29**, 1, 1976, 25-28.
[38] Representation of operators which commute with differentiation in an invariant hyperplane. — Comp. rend. Acad. bulg. Sci., **31**, 10, 1978, 1245-1248.
[39] Convolutions of right inverse operators and representation of their multipliers. — Compt. rend. Acad bulg. Sci., **31**, 11, 1978, 1377-1380.
[40] The convolutional method in operational calculus. — In: Generalized Functions and Operational Calculus. Proc. Conf. Varna, Sept. 30 — Oct. 6, 1975. Sofia, 1979, 69-88.
[41] Convolutions and approximation. — In: Constructive Function Theory '77. Proc. Internat. Conf. Blagoevgrad. Sofia, 1980, 301-306.
[42] The finite Leontiev transform: operational properties and multipliers. — Pliska. Studia Math. Bulg., **4**, 1981, 102-109.
[43] Isomorphism of the quotient fields, generated by Bessel-type differential operators. — Math. Nachr., **67**, 1975, 101-107.
[44] Convolution representation of the commutant of Gel'fond-Leont'ev integration operator. — Compt. rend. Acad. bulg. Sci., **34**, 12, 1981, 1643—1646.
[45] Lidstone-type formulas and non-harmonic sinc expansions. — In: Constructive Function Theory '81. Proc. Internat. Conf. on Constructive Function Theory. Varna, May 31 — June 6, 1981 (in print).
[46] Duhamel-type representations of solutions of non-local boundary value problems.— In: Proc. 2nd Conf. on Diff. Equations and Appls. Rousse, May 29 — June 4, 1981. Rousse, 1982, 240-247.

Dimovski, Grozdev [И. Х. Димовски, С. И. Гроздев]
[47] Бернулиево операционно смятане. — В: Математика и математическо образование. Докл. на IX пролетна конф. СМБ, Сл. бряг, 3—6 април 1980 г. София, 1980, 30-36.

Dimovski, Mineff [И. Х. Димовски, Д. М. Минев]
[48] Конволюционни представяния на решенията на гранични задачи за уравнението на топлопроводността. — В: Математика и математическо образование. Докл. на VIII пролетна конф. СМБ. Сл. бряг, 3—6 април 1979 г. София, 1979, 216-224.
[49] Convolutions, multipliers and commutants for the backward shift operator. — Pliska. Studia Math. Bulg., **4**, 1981, 128-141.
Dimovski, Petrova [И. Х. Димовски, Р. Петрова]
[50] Convolutions for a class of boundary value problems connected with the square of differentiation. — In: Математика и математическо образование. Докл. на IX пролетна конф. СМБ. Сл. бряг, 3—6 април 1980 г. София, 1980, 55-60.
Dixmier, J.
[51] Les operateurs permutables a l'operateur integral. — Portugaliae Math., **8**, 2, 1949, 73-84.
Dobrowolny, V.
[52] On fundamentals of a general operational calculus. — In: Generalized Functions and Operational Calculus. Proc. Conf. Varna, Sept. 29—Oct. 6, 1975. Sofia, 1979, 89-96.
Douglas, R. G., Shapiro, H. S., Shields, A. L.
[53] Cyclic vectors and invariant subspaces for the backward shift operator. — Ann. l'Inst. Fourier, **20**, 1, 1970, 37-76.
Edwards, R. E.
[54] Representation theorems for certain functional operators. — Pacific J. Math., **7**, 1957, 1333-1339.
[55] Functional analysis. Theory and applications. New York, 1965.
Erdelyi, A., Magnus, W., Oberhettinger, F., Tricomi, F.
[56] Higher Transcendental Functions. Vol 2. New York, 1953.
Faddeev, Faddeeva [Д. К. Фаддеев, В. Н. Фаддеева]
[57] Вычислительные методы линейной алгебры. Москва, 1963.
Gelfond, Leontiev [А. О. Гельфонд, А. Ф. Леонтьев]
[58] Об одном обобщении ряда Фурье. — Матем. сб., **29**(**71**), 1951, 477-500.
Gesztelyi, E.
[59] Generalized functions defined by fields which are isomorphic to the field of Mikusinski operators. — Publ. Math., **16**, 1969, 265-296.
Gesztelyi E., Száz A.
[60] On generalized convolution quotients. — Publ. Math., **16**, 1969, 297-305.
Glaeske, H.-J.
[61] Über die Konstruktion von Faltungen für Integraltransformationen mit orthogonalen Polynomen als Kern. — Wiss. Z. Friedrich-Schiller-Univ. Jena, Math.-Naturwiss. R., **29**. Jg., 1980, H. 2, 213-220.
Glazman, Liubich [И. М. Глазман, Ю. И. Любич]
[62] Конечномерный функциональный анализ. Москва, 1969.
Grozdev, S. I.
[63] A convolutional approach to initial value problems for equations with right invertible operators. — Compt. rend. Acad. bulg. Sci., **33**, 1, 1980, 35-38.
Hromov [А. П. Хромов]
[64] Оператор дифференцирования и ряды типа Дирихле. — Матем. заметки, **6**, 6, 1969, 759-766.
Ilyin [В. А. Ильин]
[65] Необходимые и достаточные условия базисности подсистемы собственных и присоединенных функций пучка М. В. Келдыша обыкновенных дифференциальных операторов. — ДАН СССР, **227**, 4, 1976, 796-799.
Ionkin [Н. И. Ионкин]
[66] Решение одной краевой задачи теории теплопроводности с неклассическим краевым условием. — Дифф. уравнения, **13**, 2, 1977, 294-304.
Kalish, G. K.
[67] A functional analysis proof of Titchmarsh's theorem on convolution. — J. Math. Anal. and Appls., **5**, 2, 1962, 176-183.
Kaplan, W.
[68] Operational Methods for Linear Systems. New York, 1962.
Köthe, G.
[69] Topologische lineare Räume I. Berlin-Göttingen-Heidelberg, 1960.
Krabbe, G.
[70] An algebra of generalized functions on an open interval, two-sided operational calculus. — Bull. Amer. Math. Soc., **77**, 1971, 78-84.

Krätzel, E.
[71] Bemerkungen zur Meijer-Transformation und Anwendungen. — Math. Nachr., **30**, 1965, 327-334.
Larsen, R.
[72] An Introduction to the Theory of Multipliers. Berlin-Heidelberg-New York, 1972.
Leontiev [А. Ф. Леонтьев]
[73] Ряды экспонент. Москва, 1976.
Marchenko [В. А. Марченко]
[74] Спектральная теория операторов Штурма—Лиувилля. Киев, 1972.
Máté, L.
[75] Multiplier operators and quotient algebra. — Bull. Acad. Polon. Sci., Ser. math., astr. et phys., **13**, 8, 1965, 523-526.
Meller [Н. А. Меллер]
[76] О некоторых приложениях операционного исчисления к задачам анализа. — Журн. вычисл. матем. и матем. физики, **3**, 1963, 71-78.
Mikusiński, J.
[77] Sur les fondaments du calcul opératoire. — Studia Math., **11**, 1949, 41-70.
[78] Le calcul opérationnel d'interval fini. — Studia Math., **15**, 1956, 225-251.
[79] Operational Calculus. Oxford-Warszawa, 1959.
Mikusiński, J., Ryll-Nardzewski, C.
[80] Sur le produit de composition. — Studia Math., **12**, 1952, 52-57.
Muravyev [П. А. Муравьев]
[81] Основы α-операторного исчисления. — Anal. Ştiint. Univ. "Al. I. Cuza" din Iaşi, **10**, 2, 1964, 287-302.
Nörlund, N. E.
[82] Differenzenrechnung. Berlin, 1926.
Obrechkoff [Н. Обрешков].
[83] Върху някои представяния с интеграли на реални функции. — Изв. Матем. Инст., **1**, 1, 1953, 82-110.
Povzner [А. Повзнер]
[84] О дифференциальных уравнениях типа Штурма—Лиувилля на полуоси. — Матем. сб., **23(65)**, 1, 1948, 2-52.
Przeworska-Rolewicz, D.
[85] Algebraic theory of right invertible operators. — Studia Math., **48**, 1973, 129-144.
Raichinov [И. Райчинов]
[86] О линейных операторов, перестановочных с операцией интегрирования. Математический анализ и его приложения. Ростов-на-Дону. **2**, 1970, 63-73.
[87] Linear operators defined in spaces of complex functions of many variables and commuting with the operators of integration. — Serdica. Bulg. Math. Publ., **4**, 1978, 316-323.
[88] Линейни оператори, действуващи в пространства от аналитични функции и комутиращи с фиксирана степен на оператора на интегрирането. — В : Годишник ВТУЗ. Математика, **6**, 2, 1970, 25-32.
Sebastião-e-Silva, J.
[89] As funcoes analiticas e a análise functional. — Portugal. Math., **9**, 1950, 1-130.
Shabat [Б. В. Шабат]
[90] Введение в комплексный анализ. Ч. 1—2. Москва, 1976.
Shultz, H. S.
[91] Operational calculus for functions of two variables. — SIAM J. Math. Anal., **9**, 4, 1978, 660-666.
Struble, R. A.
[92] Mikusiński operators as mappings. 1971 (preprint).
Száz, Á.
[93] Krabbe's generalized functions as convolution quotients. — Publ. Math., **19**, 1972 287-290.
[94] The multiplier extensions of admissible vector modules and the Mikusiński-type convergences. — Serdica. Bulg. Math. Publ., **3**, 1977, 82-87.
Tasche, M.
[95] Funktionalanalytische Methoden in der Operatorenrechnung. — In : Nova Acta Leopoldina, N. F., No. 231, Bd. 49. Halle (Saale), 1978.
[96] Some constructions of right inverses and generalized inverses. — In : Inverse and improperly posed problems in differential equations. Proc. Conf. on Math. and Numer. Methods. Halle (Saale), May 29 — June 2, 1979. Berlin, 1979.

[97] Operatorenrechnung in einer Algebra.—Beiträge zur Analysis, **9**, 1976, 125-130.
[98] Algebraische Operatorenrechnung für einen rechtinvertierbaren Operator.—Wiss. Z. Univ. Rostock. Math.-naturwiss. R., **23** : **9**, 1974, 735-744.
[99] Über lineare Differentialgleichungen gebrochener Ordnung und verallgemeinerte Abelsche Integralgleichungen. — Demonstratio Math., **12**, 3, 1979, 803-820.
[100] Zur Konvergenzbeschleunigung von Fourier-Reihen. — Math. Nachr., **90**, 1979, 123-134.

Titchmarsh, E. C.
[101] The zeros of certain integral functions. — Proc. London Math. Soc., **25**, 1926, 283-302.

Tkachenko [В. А. Ткаченко]
[102] Об операторах, коммутирующих с обобщенным интегрированием в пространствах аналитических функционалов. — Матем. заметки, **25**, 2, 1979, 271-282.

Volterra, V., Pérès, J.
[103] Leçons sur la composition et les fonctions pérmutables. Paris, 1924.

Weston, J. D.
[104] On the representation of operators by convolution integrals. — Pacif. J. Math., **10**, 4, 1960, 1453-1468.
[105] Time-invariant linear systems and a theorem of Titchmarsh. — In: Topics in Analysis. Colloquium on Mathematical Analysis. Jyväskylä 1970. Berlin-Heidelberg-New York, 1974, 376-383.

Whittaker, J. M.
[106] On Lidstone's series and two-point expansions of analytic functions. — Proc London Math. Soc., **36**, 1933-1934, 451-469.

AUTHOR'S INDEX

Subject Index